**Beatrice Moretti
Giorgia Tucci**

# A LANDSCAPE INFRA-STRUCTURES RESEARCH

## ROMA TUSCOLANA PILOT PROJECT

# Table of Contents

4    Preface
**LANDSCAPE INFRASTRUCTURES [AS] LAND-LINKS**
*Manuel Gausa*

8    **01/ INTERPRETATION**
STANDPOINTS AND GOALS

16    **BEYOND LANDSCAPE INFRASTRUCTURES: INNOVATION AND EXPECTATION**
*Giorgia Tucci*

24    **URBAN RENATURATIONS: GREEN CORRIDORS AND INFRASTRUCTURE**
*Nicola Canessa*

30    **LANDSCAPE INFRASTRUCTURES: MODELS OF RELATIONSHIPS IN A RESEARCH-BY-DESIGN APPROACH**
*Emanuele Sommariva*

38    **HYBRID INFRASTRUCTURES AND THE COMBINATION OF GRAY, GREEN AND BLUE: FOUR CASES**
*Matilde Pitanti*

48    **LANDSCAPE INFRASTRUCTURES CROSSOVERS: OUTLINES FOR AN ANNOTATED BIBLIOGRAPHY**
*Beatrice Moretti*

56    **THE CHERNOBYL EXCLUSION ZONE. OR INFRASTRUCTURE AS MEMORY**
*Davide Servente*

64    **GARE-MUSEE D'ORSAY. TO FILL, TO BALANCE**
*Luigi Mandraccio*

72    **TOWARDS A BACK-SIDE CULTURE. FROM ARCHITECTURAL TO PLANETARY SCALE: DESIGN ATTEMPTS TO OVERCOME THE OBSESSION OF THE FRONT.**
*Francesco Garofalo*

# Table of Contents

**82**     **02/ RESEARCH**
           READINGS AND CASE STUDIES

**88**     **INFRA/ *NATURE***
           7 case studies
**102**   **INFRA/ *PUBLIC SPACE***
           7 case studies
**116**   **INFRA/ *INDUSTRY***
           7 case studies
**130**   **INFRA/ *INFRASTRUCTURE***
           7 case studies

**146**   **03/ PILOT**
           FRAMEWORK AND METHODOLOGY
           ROMA TUSCOLANA

**158**   **9 INTERPRETATIVE MAPS**
           SOIL/
           BUILT/
           RESIDENCES/
           PUBLIC SERVICES/
           INFRASTRUCTURES/
           GREEN/
           HERITAGE/
           PUBLIC SPACES/
           BORDERS AND INTERFACES

**178**   **5 PILOT PROJECTS**
           LANDSCAPE INFRASTRUCTURE [AS]
                   *GRAFT, TRANSIT, CROSSING, BRIDGE*
           LANDSCAPE INFRASTRUCTURE [AS]
                   *HINGE, PLATFORM, INCUBATOR, GRID*

           Postface
**216**   **LANDSCAPE IS WHAT WE ARE IN**
           *Carmen Andriani*

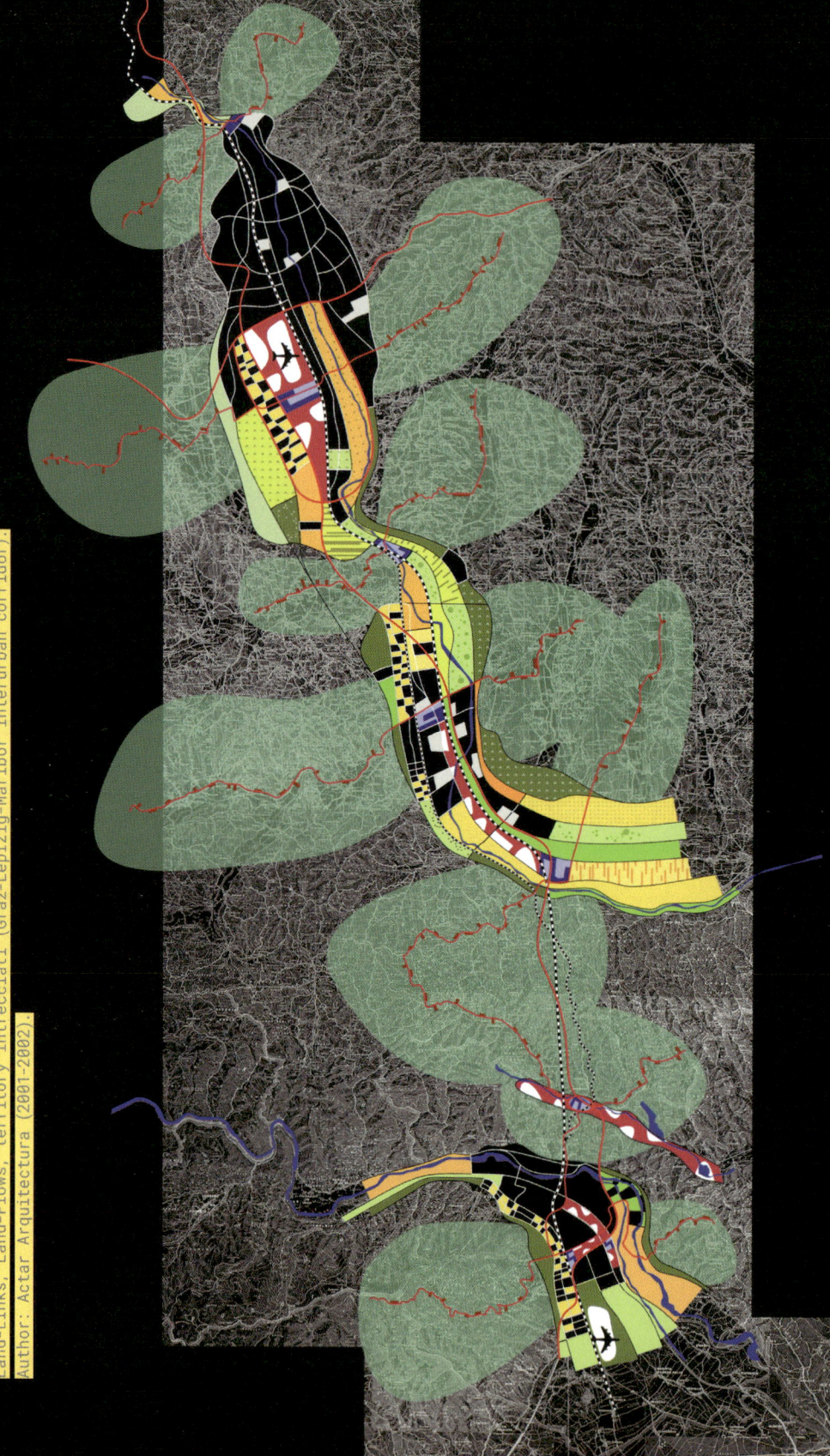

## Preface

# LANDSCAPE INFRASTRUCTURES [AS] LAND-LINKS

*Manuel Gausa*

Unlike what many of the great guardians – priests or priestesses – of the more canonical Landscaping think, the landscape is not a matter of gardening but of *gärning* (from the Swedish, action, performance, activity of a preferably global dimension). Indeed, the landscape has become, today, not only the 'other' big building but the potential 'linking' factor of the new Multi-City: the structuring element of a possible integrated and multi-scalar order – more flexible and relational – and not the 'possibilistic' remnant of the ancient *edilizia*.

A flexible device able to reset and to redirect the irregular, sprawled, random, polyhedric and, often, wild growths of our new existing urban structures and to create more versatile connective systems – not imposed but infiltrated – able to re-orientate (and/or to mesh) the metropolitan dynamics triggered today.

We have used the terms **Land-Links** to define these potential logics aiming to ensure coordinated qualitative developments, on a local and a global scale. Combining actions of coordinated developments – of 'urban systoles' and 'territorial diastoles' – in new multi-urban, relational and strategic battle-maps.

Extending and articulating landscape(s) and consolidating and reinforcing city(ies).

In this sense, the debate between the virtues of the traditional 'compact' city (density, proximity, diversity, concentration of resources) in the face of the conflicts of the more recent 'diffuse' and/or 'poly-dispersed' city (fragmentation, dissolution, disconnection, consumption of resources), is now giving way to a new paradigm, that of the plural 'intertwined' or 'interlaced' polycentric multi-city and its networked articulation. A new infra-, intra-, trans- structural and obviously eco-structural, but also info-structural model (informational) characterized by the flexible integration between systems and sub-systems, beyond the old categorical taxonomies 'city-landscape-nets' and which would, in turn, call to a new, more performative interpretation of the idea of Landscape interpreted as a 'field of forces' and 'field of relations', potentiated, reinforced and articulated in its own meta-urban structural role.

In this context the landscape, must be understood as an operational topos (a topography in its different/fluent reliefs and a topology in its differential/fluctuant geometries) capable of interacting with users and contexts and to react with/in the new informational and environmental conditions of our time.

A landscape understood not only as a 'potential' but as an active 'potency' of our contemporary 'geo-urbanities', where the architectural, infrastructural, geographic and environmental dimensions tend to meet and to intersect conditions, volitions, situations, solicitations – that is to say information(s) – in a scenario of variated (and diversified) cultures, textures, structures and natures; a scenario which calls for a new strategic and systemic dimension of this new **Cityscape** which would be now *urbs*, *natur* and *structus* at the same time.

# #1 INTER-PRETATION

«More than just steel, cement, and asphalt, infrastructure therefore forms distinctively complex, urban ecologies, a vast and immense landscape of biophysical and geospatial systems, an expansive field of resources, services, and agents that together support the landscape of contemporary economies».

- Pierre Bélanger (2017), *Landscape as Infrastructure.*

# STANDPOINTS AND GOALS

**A New Worldwide Topography**

At the turn of the 20th century, the crisis of the concept of territory as an analytical and measurable space triggered a radical change in urban studies. Cities were investigated as specific *types of territories* – e.g. conurbations, city-regions, metropolitan regions, etc. – and scholars and professionals have experienced criticalities in defining the precise nature of concepts such as urban, suburban, and rural zones. These changes gave new strategic meanings to the notion of landscape. With the digital-technological revolution, built spaces have lost a precise physical connotation and have become fields of relations. Due to the new global communication networks, there was no longer a need for territories to meet and communicate, but a growing demand for places to live and to recognize oneself. This has subjected the landscape to a process of semantic stress that, not only, has enriched its definition, but also its identity and potential applications. Invoked as a design tool, landscape has even replaced architecture as a construction model for contemporary urbanism, becoming «both the lens through which the contemporary city is represented and the medium through which it is constructed» (Waldheim, 2006).

The landscape made artificial by human activity was therefore a candidate to become the preferred means of interpreting and representing not only spaces, but also the relationships, histories and performances of places. At the end of the twentieth century, this term was used to describe all those spatial configurations resulting, for instance, from processes of decommissioning, abandonment and intensive urbanisation. Over the years, the dominance of pathways over settlements, the urban disorder, the crisis of planning and the growing illegibility of territories prompted a push to identify new references with which to interpret the transformations. They fueled the desire to comprehend the spatial forms that continued to generate, indeed defining them as "other" compared to the commonly known city. This led to the need to speak about, map out or even exhibit the landscape, or rather the new forms in which the landscape was being modeled. As urban theorists Brenner and Schimdt sharply observed, «this situation of planetary urbanisation means, paradoxically, that even spaces that lie well beyond the traditional city cores and suburban peripheries [...] have become integral parts of the worldwide urban fabric» (2011).

Already at the beginning of the 2000s, hence, we were facing the production of a *new worldwide topography*. Today, new urgent needs for energy transition combine with profound transformations in the realms of digitalization, logistics and circular economy and increase the pressure on built environments.

The challenges imposed by climate change push dynamic and diffuse systems – such as infrastructures – to equip spaces with increasingly technological and performing qualities in order to boost their resilience, fostering the coordination of multi-stakeholder and multi-institutional partnerships. Ultimately, the crisis of physicality and the relational threat triggered by the Covid-19 pandemic introduce further complexities by disrupting the relational dimension. Messages, those conveyed by gestures and physical presence, namely by the body in space, are interrupted: this requires the definition of new channels of transmission, once again immediate and decipherable.

## Etymology and Applications

In its lexical variations, the notion of landscape is capable of creating links both with the idea of determining the shape of spaces, and with a living and non-living complex of elements, which constitute the physiognomic and cultural features of a certain part of the Earth's surface. Not existing in Greek or Roman times, the term landscape first appeared in the 16th century in the Flemish artistic and literary spheres, witnessing the growing awareness of painters and writers towards what, until then, had been only the natural background of a painting or a novel.

Both the Dutch *landschap* (from which the German *Landschaft* and the English *landscape* derive) and, similarly, the French *paysage* (to which the Italian *paesaggio* and the Spanish *paisaje* are related) were born as technical terms capable of conferring an aesthetic component to nature.

If the etymology of the term landscape (in all its roots, Nordic or Latin) is particularly instructive,[1] the most interesting meaning in this context, though, is that which highlights the bond between a space and its observer. That bond that makes the landscape the product of anthropic actions, even if only of the gaze. In the Italian language, in fact, «landscape as a living mirror of our being finds its origin in the term 'country' (in Italian 'paese' n.d.a.) [...] and reminds us of the divergence between an aesthetically neutral portion of territory and the past landscape of rank, through a historical-artistic elaboration, which is anything but neutral. [...] Landscape is infrastructure because it constitutes the means by which we can enter into a relationship with the space that surrounds us, from every point of view, and which allows us to carry out activities in it. This is the task of infrastructure, that is to be in-between, and in fact the landscape is between us and nature»[2] (Kipar, 2010).

The relationship between landscape and infrastructure is subject to constant evolution, so, in this book we aim to explore, in a preliminary way, a further potential theoretical and strategic shift in this relationship. From autonomous concepts, LANDSCAPE *AND* INFRASTRUCTURE are connected through a projection in which landscape is interpreted as infrastructure (LANDSCAPE *AS* INFRASTRUCTURE). This is landscaper Pierrè Belanger's position,[3] echoed here as a foundation for reasoning. To deal with the pressing challenges that 21st century urban regions and territories are

1. In the Middle Dutch, the lemma *lantscap* combined the term *land* with the suffix *-scap*, which later evolved into the English *-ship*, indicates a widespread and overall quality or condition. In the Latin root, instead, the *pagus* was the village, hence the adjective *pagensis* meaning "the space around an agricultural village" and the verb *pango* means "to stake, to delimit and circumscribe an area". For an etymological reconstruction of the term landscape, refer to Jakob M. (2005). *Paesaggio e letteratura*. Firenze: Olschki, p. 22-23

2. Translation from Italian to English by Beatrice Moretti.

3. Reference is to Bélanger P. (2017). *Landscape as Infrastructure*. London: Routledge.

facing, Belanger introduces a new interdisciplinary perspective in which landscape architecture, civil engineering and urban planning are all aligned to achieve new theories and integrated management strategies addressing the link between landscape and infrastructure. In his studies, infrastructures are re-designed with the aim to overcome their mono-functionality and exploit them as ecological devices. Through a contamination between the two concepts, thus, this book explores potential scenarios in which infrastructures absorb and adopt the theoretical-strategic characteristics of the landscape (LANDSCAPE INFRASTRUCTURES) and, as such, become the central subject of design on the territorial and urban scale.

**Structure** The aforementioned topics are addressed in the book by means of heterogeneous contents aimed at building a-landscape-infrastructures research. The book is divided into three sections with different but coordinated objectives. The first section (INTERPRETATION) includes theoretical essays by authors with specialist skills in the fields of urbanism, architecture and landscape. The essays offer six visions of the link between landscape and infrastructure, sliding across the large scale (Canessa on urban renaturations, Sommariva about urban landscapes in transition), of the city (Pitanti on hybrid infrastructures, Servente about the idea infrastructure as memory) or of the single architectural artefact or public space (Garofalo on the innovative role of the rear in urban design, Mandraccio on the infrastructural character of public architectures). Two critical texts, written by the authors, act as annotated bibliographies on the theory and design of landscape infrastructures. By analysing crucial texts and projects from the end of the 20th century to the beginning of the new one, we aim to construct a narrative, not exhaustive but selected, of references in the field of landscape urbanism and of the design of contemporary territories. The second section (RESEARCH) captures the contemporary context by offering a selected sample of valuable case studies that address the link between infrastructure and other crucial components (nature, public space, industry) of the built environment. The third section (PILOT) reports the project results of a specific experimentation[4] of the relationship between landscape and infrastructure in the context of Roma Tuscolana railway station. The photographs showed in the book were taken in 2021 by a roman photographer Flavia Rossi under the name of "Osservatorio su un paesaggio". The project immortalises a selection of Rome's infrastructural landscapes: Roma Termini station, Porta Maggiore junction, suburban stations of Roma Casilina, Ostiense and San Pietro, Gasometro, Ponte dell'Industria and Roma Tuscolana railway junction, the priority field of project experimentation. The photographic collection develops a parallel narrative in which the landscape of Roman infrastructures is interwoven with other urban systems, offering itself as a further strategic tool for research and design.

4. The academic project, carried out in 2020 thanks to FS Sistemi Urbani S.r.l. Area Centro – Progetto Roma, was developed as a six-month topic by the "Urbanistica per il Paesaggio (Landscape Urbanism n.d.a.)" course of the Three-year Degree in Architectural Sciences (Landscape Architecture curriculum) at the Department Architecture and Design – dAD, University of Genoa (IT). The course, held by lecturers Beatrice Moretti and Giorgia Tucci, was called "LANDSCAPE INFRASTRUCTURES. Disused railway yards in Italy. The case of Roma Tuscolana".

'Hyperions', Agritectural Garden Towers For Jaypee Greens Sports City, New Delhi, India. Author: Vincent Callebaut.

# BEYOND LANDSCAPE INFRASTRUCTURES: INNOVATION AND EXPECTATION

*Giorgia Tucci*

> *Infrastructure, like technology, turns out to be a recent and promiscuous term.*
>
> *- R. Williams, 2012*

Originally, the term referred to the physical and organisational structures necessary for the functioning of a company or business. Today, it has become a much more inclusive term, assimilating to the concept of infrastructure any widely shared, human-built resource. When we speak of territorial infrastructures, we are referring to the entire spectrum of infrastructures within the territory, which originated as means/instruments for carrying out certain actions, but instead have become true identity elements of the territories they occupy. With the economic boom after the Second World War, the demographic growth and the expansion of cities, it was necessary to increase infrastructures in order to meet the needs of society, and infrastructure networks, on which the industrial economy was based, played a fundamental role.

Around the 1960s, Niles M. Hansen focused on the study of infrastructures, providing a first distinction to categorise them, dividing them into economic and social, depending on whether they affect the level of economic development of a territory directly or indirectly[1].

«The division of local public overhead capital (OC) into two components, "social" overhead capital (SOC) and "economic" overhead capital (EOC). [...] Those items classified as EOC are primarily oriented toward the support of directly productive activities or toward the movement of economic goods. SOC items [...] may also increase productivity, the way in which they do so is much less direct than in the case for EOC items» (Hansen, 1965).

Economic infrastructures (EOC), which directly support productive activities, are: roads, motorways, airports, railways, shipping, sewage networks, aqueducts, water distribution networks, gas networks, electricity networks, irrigation systems, freight transfer facilities and communication facilities. Social infrastructure (SOC), on the other hand, aimed at increasing social welfare and indirectly acting on economic productivity, includes: schools, public safety facilities, public buildings, waste disposal facilities, hospitals, sports facilities, green areas, urban renewal and rehabilitation, old people's homes, residential care facilities.

1. Cf. Hansen, Niles M. (1965), The structure and determinants of local public investment expenditures, Cambridge, Mass: MIT Press, Vol. 47.1965, 2, p. 150-162

«A region that is well equipped with infrastructure will have a comparative advantage over a region that is less well equipped, and this will translate into higher regional GDP per capita or per person in employment and/or also higher levels of employment. It follows that regional productivity, incomes and employment are an increasing function of infrastructure endowment». (Biehl, 1991). Moreover, according to Biehl, infrastructure is one of the determinants most likely to be the subject of direct intervention by economic policy makers. This latter element led to the need to quantify the real presence of infrastructures in the territory. In recent decades, however, Patrick Geddes, Lewis Mumford, Ian McHarg's work[2] has begun to consider the impacts of urbanisation of infrastructure networks on ecosystems and society, showing how infrastructures have gone beyond mere technological structures and have become shaping resources for societies, the economy and politics, demonstrating that they are not just great technical systems, but also socio-technical phenomena (Edwards, 2003). This new environmental awareness and sensitivity has led to a deeper understanding of the long-term effect of industrialisation on biophysical systems and the need for effective and comprehensive action to combat the consequences of decades of urbanisation.

For this reason, in the last two decades, ecology and urban technology experts have criticised the classical definitions of infrastructure and reasoned about an infrastructure system that is more compatible with post-industrial societies. Edward McMahon and Mark Benedict, planners in the field of sustainability, advocated the introduction of 'green infrastructure', as opposed to Grey Infrastructure such as roads, sewers and utility lines or Social Infrastructure including hospitals, schools and prisons, emphasising the importance of separating natural and man-made infrastructure in planning and development approaches [3].

This new category of infrastructures - green and blue infrastructures - which can be assimilated to both economic and social infrastructures, as they provide ecological, economic and social benefits through environmental enhancement, has rapidly found its place on theoretical and political levels. The European Commission has defined green infrastructure as "a network of natural and semi-natural areas strategically planned with other environmental elements, designed and managed to provide a wide range of ecosystem services. This includes green spaces (or blue spaces, in the case of aquatic ecosystems) and other physical elements in areas on land (including coastal areas) and at sea. On land, green infrastructure is present in a rural and urban context" (EEA, 2011) [4]. Green infrastructures are based on the need to protect and improve nature and natural processes, as well as provide multiple benefits to society, and should be consciously integrated into spatial planning and development programmes, since compared to traditional infrastructures (also called grey infrastructures), which are mainly designed with a single purpose in mind, green infrastructures have multiple benefits (improves

2. Reference to concepts expressed in studies: Geddes, P. - Civics as applied sociology (1905); Civics: As concrete and applied sociology (1906); Cities in Evolution: An Introduction to the Town Planning Movement and the Study of Civics (1915); Mumford, L. - Technics and Civilization (1934); The Culture of Cities (1938); The Condition of Man (1944), and The Conduct of Life (1951); McHarg, I. - Design with Nature (1969); A Quest for Life (1996)

3. Cf. McMahon, M. B. E. (2002). Green Infrastructure: Smart Conservation for the 21st Century. Renewable Resources Journal, (20): 12-17; Benedict, M.A., McMahon, E.T. (2006), Green Infrastructure: Linking Landscapes and Communities, Arlington, Virginia: The Conservation Fund

4. Cf. European Environment Agency (EEA), Green Infrastructure and territorial cohesion. The concept of green infrastructure and its integration into policies using monitoring systems. EEA Technical report No 18/2011

water and air quality, fights social exclusion, improves the quality of spaces, encourages aggregation, mitigates climate change, reduces urban heat islands…). Green infrastructures represent a significant renewal in the dynamics of territorial development, providing an alternative to traditional "grey" solutions, which have contributed to the profound changes to which the earth's ecosystem has been and continues to be subjected with serious environmental consequences. To combat the loss of land, the exploitation of environmental resources, increasing global warming and the loss of biodiversity, global policies in recent decades have been committed to pursuing common goals to cope with what scientists believe will be the sixth mass extinction, i.e. the next biotic transition within a geological era in which the Earth's ecosystem undergoes a profound transformation, triggering an extinction process. Since the 80s, the scientific community has shared the view that the world has entered a new geological era, known as the Anthropocene [5], in which the activities of human beings are significantly and irreversibly changing land structures, ecosystems and the climate on the planet. Human activity has altered between 50% and 75% of the earth's surface in an attempt to create space for cultivated fields, cities and infrastructure, cementing natural areas, promoting soil erosion, destroying biodiversity and polluting the atmosphere, thus dominating 90% of the ecosystems.

This complex global context has led to a significant change in thinking about reinterpreting infrastructure as a strategic tool for contributing to environmental and social resilience. All over the world, virtuous recovery processes of disused infrastructures have been activated in order to plan, program and implement connected networks of natural and semi-natural areas (e.g. agricultural or peri-urban areas), to stop degradation and ensure, maintain and develop a series of ecosystem services inside and outside cities.
There are numerous green infrastructure projects completed or under construction around the world, from the UK's Green Belts, to Barcelona's Anella Verda or Territorial Planning in the Lisbon metropolitan area, as well as large urban green infrastructure projects in the United States, an area often compromised by climatic phenomena. The same applies to blue infrastructure projects, whether for marine environments or inland waters. Along river courses, as well as acting as an ecological corridor to improve the integrity of the ecosystem, they can be designed to renaturalise flood control areas and restore peripheral wetlands, and within cities to play a valuable role in regulating urban heat islands. In the marine environment, for example, Posidonia oceanica meadows have multiple roles: they protect biodiversity as nursery areas, they combat coastal erosion, they are important for regulating oxygen, storing carbon and capturing CO2, even tens of times faster than terrestrial vegetation. Canals become 'living filters' to combat flooding (LILA 2020: Phase Shifts Park, Taiwan, 2019; Weiliu Wetland Park, China, 2017), streets are transformed into multifunctional spaces (Vestenpark Hendrik Speecqvest, Belgium, 2019; Rambla de

5. "The 'Anthropocene' is a term widely used since its coining by Paul Crutzen and Eugene Stoermer in 2000 to denote the present geological time interval, in which many conditions and processes on Earth are profoundly altered by human impact. This impact has intensified significantly since the onset of industrialization, taking us out of the Earth System state typical of the Holocene Epoch that post-dates the last glaciation" - Working Group on the 'Anthropocene' | Subcommission on Quaternary Stratigraphy, *quaternary.stratigraphy.org*

LILA 2020: Phase Shifts Park, Taiwan, 2019 by Mosbach Paysagistes.

Vestenpark Hendrik Speecqvest, Belgium, Mechelen 2019 by OMGEVING Landscape Architecture.

Wuhan Yangtze Riverfront Park, China, 2018 by Sasaki.

Empire Stores, New York City, USA, 2016 by Future Green Studio.

Vancouver Waterfront Park, Portland, Washington, 2019 by PWL Partnership Landscape Architects.

Grand Park: The Park for Everyone, California, Los Angeles, 2012 by RIOS.

Sants, Barcelona, 2016), flood channels into natural spaces (Lower Factory Pond, Switzerland, 2011; Wuhan Yangtze Riverfront Park, China, 2018), disused railway infrastructure into recreational spaces and slow mobility connections (Albany Loop, California, 2019; Park am Gleisdreieck, Berlin, 2014), former industrial districts into parks for biodiversity (Empire Stores, USA, 2016), unused airports into public sport-cultural spaces (Tempelhof Feld Park, Berlin, 2010), waterfronts into aggregative spaces (Vancouver Waterfront Park, USA, 2019), parking plates into public parks for children (Grand Park, Los Angeles, 2012; The Beach at Expedia Group, Seattle, 2019), etc.

The benefits that the introduction of this new green infrastructure has brought are visible and proven globally. Within urban areas, they bring health benefits, such as clean air and better quality of water, strengthen the sense of community, consolidate links with voluntary actions promoted by civil society and help to counter exclusion and social isolation, benefiting citizens and the community physically, psychologically, emotionally and socio-economically. Urban gardens and community gardens also promote urban food production by reducing the distance between production and consumption. Green solutions approaches have helped urbanised areas to counteract the negative effects of climate change, such as flooding and heat islands - see the strategies implemented in China to mitigate disasters caused by dense urban construction leading to severe flooding. Finally, green and blue infrastructure can protect natural habitats and promote biodiversity, contributing significantly to global goals.

However, green infrastructure development is at a crossroads. "Over the past 20 years, an increasing number of green infrastructure projects have been implemented, backed by a wealth of empirical data showing that this approach is flexible, robust and effective. Green infrastructure projects have been implemented at local, regional, national or cross-border level. However, in order to optimise their operation and maximise their potential, interventions at different levels should be interconnected and interdependent. This means that benefits are greatly enhanced if a minimum level of relevance and coherence between the different levels is achieved. If no action is taken at EU level, only a few independent initiatives will be implemented, which will not maximise their potential to restore natural capital and cut the costs of heavy infrastructure" (EU, 2012)[6]. Geographical elements such as mountain ranges (Alps, Pyrenees, Carpathians), river basins (Rhine, Danube) and forests (Finnish-Scandinavian forests), which extend beyond national borders and are part of the shared natural and cultural heritage, require joint actions and a pan-European vision.

According to EU recommendations[7], it is therefore essential to use green infrastructure in spatial planning and development processes and to integrate it fully into the implementation of not only regional but also trans-European policies. The potential of green infrastructure will be further enhanced by ongoing technological development and the promotion of the bio-economy. Through

6. Cf. European Commission, Bruxelles, 2013. Infrastrutture verdi - Rafforzare il capitale naturale in Europa, p.9

7. European Commission, Science for Environment Policy, In-depth Reports. The Multifunctionality of Green Infrastructure, March 2012

appropriate processes, significant improvements can be achieved in particular fields like: transport, energy, agriculture, urban design and management. In the future, green infrastructures may be fully absorbed by cities through resource-efficient and energy-efficient 'smart' buildings with roof gardens, green walls and innovative materials, providing environmental, social and health benefits. The integration of sensors, Big Data and Artificial Intelligence in the design of cities offers the possibility of constantly monitoring urban centres, helping to manage emergencies such as potentially critical environmental phenomena, communication and mobility systems, services and infrastructures, making it possible to develop strategies and solutions that start from a specific analysis of contexts useful for identifying the problems and resources present in the area at infrastructural, economic, sociological, cultural and technological levels.

This new vision for the city aims to improve the quality of citizens life, but also to build systems of urban experience that are not just made up, but co-designed with users and citizens, increasing the sense of participation and belonging by creating visions of a city that truly adapts characteristics and desires of population. The green infrastructure merges with the urban landscape with the aim of "reconnecting people with nature" - LAND research[8] - declining the strategy of regenerative urban development and considering the urban context as a shared ecosystem working with nature to provide environmental, social and economic benefits to people. The new urban-green-digital landscape will work in an interconnected way, measuring sustainability performance, providing constant monitoring and increasing environmental awareness. This innovative vision has already found several prototyping opportunities in pioneering projects for the cities of the future, such as the Woven City in Higashi-Fuji which aims to create a model of an automated city and solve social problems, or the Europa City in Paris, the Hyperions project in India, the Neom city project in Saudi Arabia, etc. There is a lot of expectation for the future, an expectation that today is getting closer and closer to reality as technological experimentation has reached avant-garde levels and a collective awareness has spread that it is no longer possible to continue living in consumerist and impactful cities. The grey, green or blue infrastructures, hitherto understood as separate entities, will have to adapt to the new economic-urban dynamics and the new socio-environmental needs, becoming a combination of technological innovation and environmental sustainability.

8. Cf. Landscape Architecture Nature Development, LAND Design Research *www.landsrl.com/ adaptivedesign*

Woven City, the first techno-city of the future built. Authors: Toyota in Higashi, Fuji and designed by the BIG studio.

# URBAN RENATURATIONS: GREEN CORRIDORS AND INFRASTRUCTURE

*Nicola Canessa*

More and more in the last decades the relationship between the concepts of space, culture and movement has been growing, the same ideas of dimension and time require new cities able to absorb and be absorbed by the people who live and travel in them. Talking about cities today means talking about an organism capable of connecting on a local and global level with increasingly differentiated and specialized people and/or users who seek new references, stimuli and experiences in the territory. A new city should be able to transform its history into a new interpretation of space, in an unprecedented way. With this attitude, today's users can reconfigure and treat the territory as more than just a series of events and think of it as a series of experiences.

The reconfiguration of urban structures in the last century has become the source of urban development itself, just as the transformation of abandoned industrial areas is usually linked to the reconquest of space by civil society, the transformation of large travel infrastructure networks, railways, highways, freeways and city grids, towards a reconfiguration into new public spaces, coupled with sustainable transport, can also become an interesting new urban proposal.

Today, urban renewal and reconstruction are important issues on the agenda of all major cities around the world. In recent years, there has been a trend towards the restoration of disused structures (such as abandoned railway lines) as traces of memory and parts of the industrial economy that are now obsolete. Even the Italian Railways have realized the need to reconstruct these spaces, rich in potential, and the company has recently released an atlas of abandoned lines. The purpose of the atlas is to allow interested associations, public administration agencies and competent authorities to identify and evaluate the many opportunities offered by abandoned routes or spaces.

This is because in the rest of the world there is already a tradition, we might say the recovery of these artifacts, for example the Promenade Plantée in Paris is the first elevated linear park in the world, built in the late 80s and inaugurated in the 90s, born on an abandoned railway line that since 1859 connects Place de la Bastille to Varenne-Saint-Maur. The example in turn inspired the creation of another well-known linear green space: the High Line Park, a beautiful garden in New York City, or a city park, resulting from the conversion of the historic freight line over Manhattan's West Side. For decades an abandoned area, today we find that flowers, gardens, railroad tracks, trees and convivial spaces harmoniously compose themselves in the post-industrial reality of the metropolis. The Municipality of Milan is also developing stimulus projects in this direction. On October 19, 2016, the "Green Track" project took

Model of the project MultiRamblas & MultiEnsanche: new centralities, realized for the exhibition "Re-cycle, strategie per l'architettura, la città e il pianeta" (dir: P. Ciorra, S. Marini) at MAXXI, Museo Nazionale delle Arti del XXI Secolo, Rome, November 2011-April 2012. Designer: Gausa+Raveau actarquitectura, with Laboratorio GIC-Lab. UNIGE-Genova (Barcelona Ensanche-Multi-String New Centrality, Barcelona, 2010), with Intelligent Coast group (Barcelona Multiramblas, Barcelona, 2009).

place in the center of Milan. With the support of Fondazione Cariplo, WWF developed the project in collaboration with the Municipality of Milan, Cooperativa Eliante and Rete Ferroviaria Italiana, with the aim of transforming abandoned railway yards into green oases. The aim of the pilot project was to conduct a feasibility study to use the operating railway buffer zone as a connecting element, creating a real linear park between San Cristoforo and Porta Romana. The aim of the project is to create a structure that penetrates the matrix of urbanization, providing the city with a new concept of urban park, characterized by a "green field". This idea of being able to inject a new green plot within the cities by taking the cues from urban structures that are no longer used or possibly not usable given the possibility of a relief in local private transportation, is very interesting.

When we think of nature, we think of something uncontaminated, sometimes even a little wild. A space of discovery and also of rediscovery of ourselves, where we go to recharge our energies. Even entering a forest of chestnut trees, we feel as if we are becoming explorers again, discovering new places never explored before. In reality, we are following in the footsteps of someone else, often that forest has been planted and made to grow by someone in order to recover the fruits in a controlled manner. We consider incredible and rightly worthy of protection environments such as the Cinque Terre, which are an emblem of human anthropization, the ability to shape the territory in order to make the most of its characteristics and survive within it, from the houses overlooking the sea and terraces in the mountains you have to feel the tunnels that cross everything is part of a wise game of give and take between nature and man. The parks of our cities are spaces that have been shaped to enjoy the presence of nature within the built environment, but they are more or less all an artifice. Take Genoa, its two parks in the city center are one built on a quarry of stonemasons inside the city, then born from a process of renaturalization by Barabino, while the other Baltimore Gardens also called Plastic Gardens, are the result of a short-sighted reconstruction, but still fascinating, after the demolition of an entire district of the historic center to make room for a new business center that has never really taken off and where the park is located on a concrete slab that covers the junction between urban and interurban roads. The natures of our cities are therefore very often "unnatural" (Canessa, 2021). This being unnatural, does not make them less interesting or less functional, on the contrary sometimes makes them more suitable for man, even in large urban renaturalization projects, we find today even more than yesterday the ability to create multifunctional landscapes able to make not only our cities more livable spaces, but also and above all places lived by people, creating convivial and sometimes even resilient spaces (Gausa, Canessa, 2020).

The new green infrastructures that are springing up in cities all over the world are new tools to bring man closer to nature, they are spaces that create new urban complexity and new quality, but they are also memory tools. In this cultural context, our research work as GIClab sees cities as natural laboratories of new paradigms

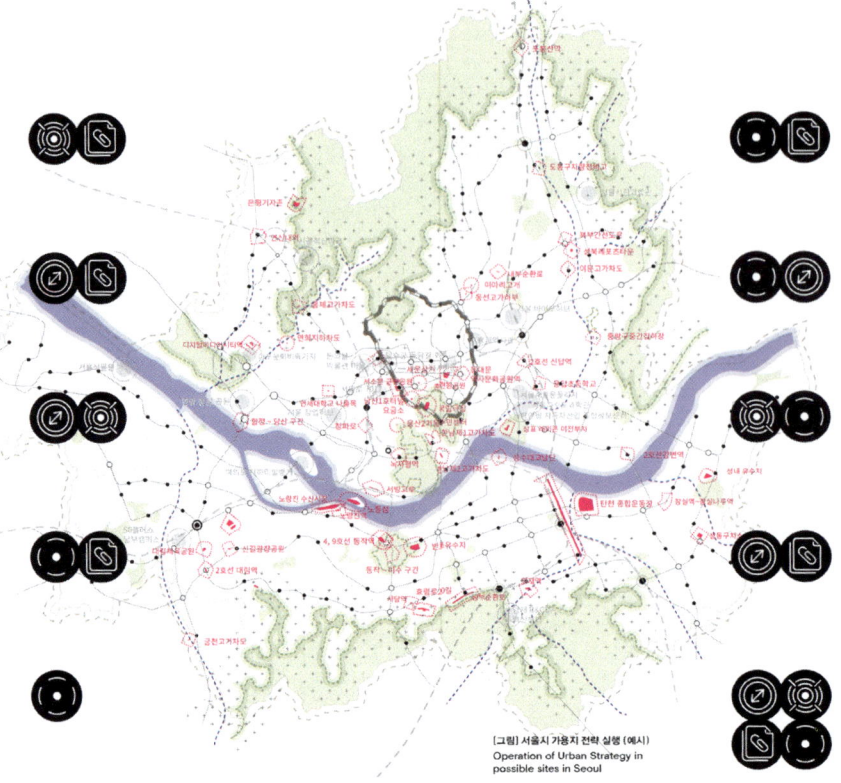

[그림] 서울시 가용지 전략 실행 (예시)
Operation of Urban Strategy in possible sites in Seoul

for new dynamics, processes and current needs triggered and introduced by the ever increasing demand for change in urban quality. In the context of our research, the Mediterranean port-cities so intrinsically predisposed and implicitly devoted to this relational dialectic are configured as privileged theaters and amplified logical and systemic evolution of a complex space of new conception. Landscape, environment, development in the articulation of these territories, in fact, are found more than in any other reality to be confronted and mediated in and between permanence, stratification and interweaving, in a concert of pushes and resistances to change, with respect to which projects such as New Multistring City (Gausa, 2012), although developed in a specific context such as Barcelona, are proposed as strategic paradigms in the definition of new tactics in the face of the need to combine and resolve tensions and questions of the great themes of today's debate. Or in Seoul where in 2018 the City Council decided to launch an international tender for the development of guidelines for the renaturalization of urban infrastructure, won by Gausa, Go-Up + GiClab and Lokaldesign with the project City Civic Seoul (Shin, Gausa, Canessa, 2019), which aimed to revise the use of major roads in view of a process of reduction of circulating vehicles and the implementation of public and electric transport, with a large plan of planting and urban transformation, aimed at creating new public routes and convivial spaces, The project then led the municipality to launch an

Map of strategies for the renaturalization of the city of Seoul for the Civic City Seoul project.
Authors: Manuel Gausa, Go-Up + GiClab and Lokaldesign (2018).

Underground competition and a Superground public consultation to investigate in more detail some clusters of transformation.
In Genoa, too, we have been working with GIClab on processes of renaturalization of urban grids or disused industrial structures, and it is interesting to see how here, as in Barcelona with the Sagrera park or in Milan with the project for disused docks, the perception of the renaturalization of disused infrastructural axes has been developing for some years now. One example is the disused route in Valpolcevera, the route of the former railway line towards the protected area of Madonna della Guardia. Today this part has been abandoned for decades and has been transformed into a linear park path where the switches and tunnels that cross the forest up to the height of Mount Figogna are still clearly visible. This recovery project was carried out by the small town of Ceranesi, which cleaned and protected the exposed part and transformed it into a hiking and biking trail of particular beauty.

In today's dimension where "the ancient geographical borders, aimed at containing the new emerging city, have given way, almost suddenly, in front of the different scales of a new field of action, much more complex, elusive and vital, in which latent nuclei and consolidated nodes, uncertain margins and spaces of friction, consolidated fabrics and unfinished plots coexist, thus announcing the new condition mestizo and progressively ambivalent (between the natural and the artificial) of a new territorial urban scenario" (Gausa, 2009) what characterizes and distinguishes the approach proposed with New Multistring City is the willingness and possibility to combine, in the same project, simultaneously and synergistically reconversion, recovery, renaturalization, but also innovation, interconnection and hybridization, multiplication in the definition of strategies and space renacturativate with respect to which spaces and territories become a platform for the autonomous choice of the user who can at any time not only choose and modify, but even draw and define, for their own use and consumption, according to their personal and intimate aspirations and concerns, through their own actions and feelings, new urban geographies without putting in crisis the overall system.

### Bibliography

Canessa N. (2021), Unnatural. Territorial relation between city and nature, Actar, New York.

Canessa N. (2020), Resesili(G)Ence Vol.2. GOA Resili(g)ent City, Actar, New York.

Gausa M. (2020), Resesili(G)Ence. Vo.1. Intelligent Cities / Resilient Landscapes, Actar, New York.

Gausa M. (2012), BCN-GOA New Multistring City, List Laboratorio Internazionale Editoriale, Trento.

Gausa M. (2009), Multi-Barcelona. Hiper-Catalunya, List, Trento/Barcelona, 2009.

Shin H., Gausa M., Canessa N. (2018). I.Seoul.U: Basic Survey for the Urban Visions of Civic City Seoul. vol. 1, Genova:Textures.

Dömitz former railway bridge over Elbe river in Wendland. Photo: Emanuele Sommariva (2018).

# LANDSCAPE INFRASTRUCTURES: MODELS OF RELATIONSHIPS IN A RESEARCH-BY-DESIGN APPROACH

*Emanuele Sommariva*

> *The identity continues to be true whatever the value of the symbols of which it is composed, impling continue adaptations...*
> *In the same way, the uniqueness of each territory cannot be permanent, but it araises from cycles of changes and transformations.*
>
> *- G. De Carlo, 1998*

**Urban Landscapes in transition: design methodological premises**

It's with these words in defence of the design thinking approach, that I would like to recall the multifaceted work of Giancarlo De Carlo —in the course of the events organized the centennial of his birth— regarding the importance of design culture as a tool for prefiguring collective needs in response to the evolution of Space and Society. De Carlo, who many times during his intellectual and work experience came back to question himself on the value of the "public domain" as a fundamental category of knowledge for the governance of transformations, underlines to overcome pure formalism and visions promoted through the mega-structuralism of the Modern movement, as described by Tafuri in the International of Utopia (1976). Principles which referred to a prodrome of the "system theory" approach applied to design disciplines and necessary for breaking disciplinary barriers through "radical thinking". Term, coined by Grabow and Heskin (1973), as the first theories of participatory urban planning, expressed by figures such as G. De Carlo, Y. Friedmann, L. Kroll and B. Rudofsky, who understood the project no more as an end point, but as a tool through which alternative forms of habitability and structural challenges could be tested. A lesson the one by De Carlo which is still timely: questioning on the structural gaps of the contemporary urban complexity, on the one hand, and on the crisis of design disciplines as traditionally understood, on the other. A J'accuse which urges us to escape from the degeneration of failed utopias no longer able to govern the formal quality of both the physical interventions and the spatial relationships of the post-metropolis, as well as to envision the future of our territories in a dimension of material scarcity. Thus, in the era of the Global city the conceptual superimposition of landscape as infrastructure (Allen, 1999; Bélanger, 2016) become evident through the predominant challenges facing urban regions and territories today —including shifting climates, material flows, ecological performances and population mobilities— while patterns of spaces produced by sub-urbanization processes contrast in size the scale of the consolidated cities. Palimpsests (Corboz, 1983) of a human-dominated geological epoch: the Antropocene, (Sijmons, 2014) in which the elusive game of local identity is incessantly rewritten, between places that have never been completely forgotten and non-places (Augè, 1992) that are not completely fulfilled. As such, the architecture of the urban landscapes has evolved into a complex system, in favour of economic growth and the development of nations as primary field of investment of public/private authorities (Shannon, Smets, 2010), by overcoming the relational ontology of "humanity-in-nature" and the dialectic dualism of "nature and society". In this specific context, the understanding of the interactions between physical environments and infrastructures plays a key-role providing the pre-condition of any spatial development and to support socio-ecological dynamics. In other terms, infrastructures, by virtue of their scale, ubiquity and inability to be hidden, are an essential component of the urban landscape (Strang, 1996). Without discussing the different trajectories in conceptualizing the operative definitions of the terms, in this context Landscape Infrastructures can be defined as "constructed facilities and natural assets that shelter and support most urban/rural activities —production and distribution facilities, public fixed capitals and amenities, ICT and welfare services, transportation and mobility, materials supply and natural cycles— whose organization, management and characteristics is the result of the action and interaction of human and environmental factors". The opportunity to redefine both notions into a more integral framework, where cross-cutting topics converge, contribute to the debate on the repositioning of the design culture, as already sustained by De Carlo with ILAUD international workshops (Occhialini, 2005). Landscape become an operative field where infra-structures deploys functions and narratives form of the architecture of the territory, in order to sustain urban processes and social organizational bases. Therefore, this paper aims to explore the concept of landscape infrastructure in research-by-design approach, but to re-establish the role of design as integrating practice at multiple scales of intervention. This implies interdisciplinary fields of investigation, where the principles of multi-functionality, connectivity, integration, long term sustainability, ecology, social-inclusivity are at the core of spatial design solutions. Furthermore, the paper outlines a set of research strategies for landscape infrastructures focusing on four potential fields of applicability.

The idea to conceive landscape infrastructures as medium to shape the architecture of territories (Schröder, Carta, 2017) is a long-lasting debate, which includes in its broadest sense other design categories, such as the confront of indoor and outdoor spaces, flows and patterns logics, local conditions and regional-oriented networks. In fact, the architecture of territories has evolved into a complex mechanism extending deep into the earth and far into the rural landscapes (Strang, 1996), beyond any direct perception of how much infrastructures serve as support for shaping tangible relationship in a region and to establish local identity. At the same time, a given landscape confronts the designer with a series of challenges and opportunities that must be addressed with clarity not only to create new layers or urban landmarks, but also to embody architectural work with common-sense, functional needs and cultural values. The continuous dialogue among architecture and territory reveals significant modes of relationships, among others

**From Land-Arch approach to Operative Landscapes interpretation**

the principles of contrast, merger, and reciprocity are useful categories to read the significance of a project and to embody a rich complexity of meanings.

## I. Contrast > nature apart
*or the act of juxtaposing architecture with the landscape as in a scenery*

Ictinus + Callicrate _ **Parthenon**
Acropolis of Athens, 447 - 432 BC

Architectural work stands apart from a given context or cultural landscape in terms of volumes, materials, profiles, scale, in concert to create a powerful counterpoint to its immediate setting with no transitions. Landscape is seen as autonomous with its own processes, ecosystems and qualities, and generally unaffected by the architect. The application of this principle can be traced throughout history of architecture, from Athens' Parthenon (Ictinus + Callicrate, 447-432 B.C.) standing as the core element of the Greek Acropolis, to New York's Central Park (F.L. Olmsted, 1857-76), whose sequences of pastoral and picturesque scenery constituted a totally different environment from the surrounding urban grid. Even if apparently so different, the two examples share the concept to provide an experience of nature apart understood in Ruskinian and Emersonian terms: a powerful piece of environmental sculpture (not different from the marble carved for the Parthenon) where the surrounding landscape represent the structural scaenae frons to nurtured the cultural and moral well-being of the citizens.

Frederick Law Olmsted _ **Central Park**
New York, 1857 - 1876

## II. Merger > mimicking nature
*or the act of integrating architecture with the landscape according to the surrounding qualities*

Frank Lloyd Wright _ **Fallingwater**
Mill Run US, 1936 - 1939

As opposite of contrast, the project is designed to appear as an integral part of its context, in terms of volumes, materials, profiles, scale, or reflecting qualities of the surrounding landscape conditions, by implying rational-scientific attitude toward nature, as a complex realm of dynamic processes to be respected. Applied strategies related to this concept can be found in much of Wright's work, especially Taliesen West and Fallingwater (1936-39). Landscaping become a design tool to adapt projects to topographic, climatic or site-specific conditions mimicking nature as a structural pre-condition. Same approach related to landscape architecture works can be found in Richard Haag's sequence of forest gardens in Bloedel Reserve (1970-87), where the introduction of few elements, such as hedges, stone gardens, reflection ponds, mosses and ferns highlight the formal design of clearings, while evoking the transition of an unaltered bog merged with topics of Japanese garden.

Richard Haag _ **Bloedel Reserve**
Bainbridge Island Seattle US, 1970 - 1987

## III. Reciprocity > intermediate nature
*or the act of mutual influencing architecture with the landscape in a complex new whole*

Is the most frequently employed relational concept, where architectural work and landscape are seen as artifacts in continuous dialogue, modifying each other according to organizational structuring components, indoor/outdoor spaces, volumes/surfaces, materials/forms, topography/views, accessibility/intimacy. The views of intermediate nature that inform reciprocity are as varied as the formal strategies used to express it. Among these is worthy to mention the work of Pirro Ligorio at Villa d'Este in Tivoli, Rome (1550-1605), where extraordinary system of fountains, water games, nymphaeum and a labyrinthic garden has been designed, by diverting part of nearby Aniene River and prompting a spectacular view over the Ager Romanus. In order ways, reciprocity epitomizes Eero Saarinen and Dan Kiley modernist approach for the design of Miller House (1953). With its flat roof and flowing layout, rotating around the central living room, the correspondence with the outer garden has been conceived as an extension of the house; avoiding conventional axial organization, Kiley loosely divided the open space into three sections, extending from the corresponding partition of the house, each with its own identity. One of the most interesting aspects of the present scene in the urban design debate is the wide variety of conceptual strategies in an attempt to codify a research-based culture of dialogue on the architecture of territory, according to different key-reading features, such as: visibility (context); livelihood (networking); performativity (ecology); multifunctionality (density); inter-scalability (layering); synergies (programmes); identity (culture); variability (potentials); adaptability (resilience); replicability (logics). This requires overcoming the classical ontological relationship based on traditional figure-background categories (building/ground, city/countryside, territory/landscape), where also the guiding principles of contrast, merger, and reciprocity are enriched of new, open and fuzzy interpretations. New geographies of transitions (Gausa et al., 2003) where the application of new structural logics and technical concepts address the operative condition of landscapes. Today the strength of the bionomy of Landscape and Infrastructure resides precisely in their capability of evoking vivid images of multiple natures —evidently dramatic, mongrel, manipulated, rather than domestic and bucolic— that cannot be handled with traditional planning approach. The territorial forms of continuity and changes are proposed to become the real contexts of local communities, while a more comprehensive systemic vision is necessary (Mostafavi, Doherthy, 2011).

Pirro Ligorio_ **Villa d'Este**
Tivoli Rome, 1550 - 1605

Eero Saarinen + Dan Kiley _ **Miller House**
Columbus US, 1953

The demand for environmental sensitivity in the era of digital knowledge, reaffirm the necessity of socio-spatial innovation as driver of change, able to shape the future of our cities and the architecture of territories. By implementing the advantages offered by the digital milieu and hyper-medial ecosystems (e.g., internet of things, just-in-time logistics, E-commerce, user profiling, big data), infrastructural landscape point new architecture of aggregates: systems of environmental, physical or immaterial relationships; mutation and morphoses of adaptive

```
Infra-scapes
visions:
new logics
```

Agnes Denes _ **Wheatfield**
New York, 1982

Studio Landraum _ **Agropolis**
Munich, 2009

Field Operations _ **Fresh Kills Park**
Staten Island NY, 2001

renewal cycles of territories (Vogt, Kissling, 2020); counterparts of the universe of natural dynamics in which we live. Although the interest in landscape urbanism (Waldheim, 2016) is taking up the contemporary debate on landscape-oriented and systemic-design strategies, the terms "Recycling", "Regeneration" "Recovery", "Re-naturalization" identify further more fields of action-research and challenges for current urban and territorial agendas, to respond to the demand for the Open City of the future argued by Richard Sennet (2018). Planning the new urban condition, in fact, it means thinking not only the spatial form of the cities, which is always subject to change, but to conceive its forms of aggregation on different semantic levels: the asymmetric formulation of markets' offer/demand, the extensive re-production of places for distribution/consumption, the physical/digital infrastructural imbalances, the evolution of consumers trends, as well as the exploration of new habits, rituals and job opportunities, which are influenced by processes of creative disruptiveness. In the same way that contemporary city superimposes layers of information in successive and varied entropic combinations, the project can also superimpose simultaneous horizontal field conditions —in terms of interconnected programmes, morphologies, structures, uses— building integrated resilience in Social-Ecological Systems in a vision of sustainability, as persistence borne out of change, multiplication and redundancy. Many of the most promising ideas in this regard are the reformulation and reclamation of the in-betweens: partitioning of open spaces and articulation of clustered activities, as real generators of urban life, planning towards new integrated land-uses and patterns. Four interpretative strategies have been identified to further investigate the various aspects of these phenomena in the contemporary design practices:

### *Strategy 1. Eco-structures: surfaces as volume*

Landscape, at the very end, is an art of surface; extending the traditional topographic articulation has become a primary instrument in design. The manipulation of surfaces has been always a constant for architecture, transforming an element that usually bears a flat coding into an active, complex, mutating field. The ambiguity between the surfaces (2D) and the production of spaces (3D) is no longer related to a precise limit, land-use or static condition; is determined by the enabling conditions it can produce in place. Understanding the complex scales and dynamics of adaption and change of land matrices can be described as the interaction of flexible structure, dense spaces, articulated streams, drosscapes and new surface conditions. New concepts of land-links, land grids and eco-streams (Gausa et al., 2003) become paradigmatic of a conceptual approach to landscape infrastructures. Examples in this regard can be found in the land-art approach by Agnes Denes' with the Wheatfield (1982) installation at Battery Park, New York; or the radical scenario approach envisioned by Studio LandRaum for for Agropolis Munich: an urban productive landscape for Munich metropolitan area (Schröder et al., 2009).

### Strategy 2. Land-recovery: voids as potential

As entropy increases, the critical mass of the contradictory accumulation of information, uses, programme and former structures grows over time, confronting the traditional notion of land ownership in favour of temporary, movable, flexible organizations, re-informing cities though weak urbanization models (Branzi, 2006). What remains out of the process of fractal growth of urban structures become a latent void, affecting not only the Edge-Cities (Garrau, 1991), but also recursive form of peripheral conditions and shirnkage, which also affect inner or consolidated part of urban areas, in a successive positive-negative, or empty-full seriations. Processual transformation and re-cycling of vacant, abandoned and wasted urban sites, former industrial areas, and territorial assets figures to be one of the great infrastructural design challenges, by implementing natural remediation approach and environmental engineering techniques, such as Peter Latz's IBA Emsher Park at Duisburg Nord (1990-2002); the Fresh Kills landfill's environmental remediation (2001) by Field Operation; or again NY's High Line designed by Diller Scofidio+Renfro (2004-09).

Field Operations + Diller&Scofidio _ **High Line**
New York US, 2004 - 2009

### Strategy 3. Tactical Co-design: performances as process

The rise of interest in planning and policymaking on the temporary nature of tactical and co-designed pilot projects, allowed in recent years to observe interventions and acquire knowledge on the ground before committing to long-term transformations. Design processes in this way do not focus primarily on isolated question of forms, but on sharing responsibilities, practicing everyday life changes and civic call-to-action. Short-term commitments and realistic expectations to facilitate consolidation, evolution and proliferation of a set of dynamics and rules are at the base of tactical urbanism approach (Lydon, Garcia, 2015). It is another interpretation of enabling technologies (info-structures) shifting the focus from essence to effects produced in place. Investigating the feed-back loops between architecture and the context, tactical co-design approach does not renounce to formal aspects, but builds it as maps of actions and agopunctures, such as in Metrocable project (2007-10) by Urban Think Thank which implements a safe mobility infrastructure, in Barrio San Agustin, Caracars; or inspiring D.I.Y tactics and guerrilla gardening campaigns to raise civic care of derelict open spaces such as in Jardin DeMAIN (2010) by Atelier Coloco, in Montpellier.

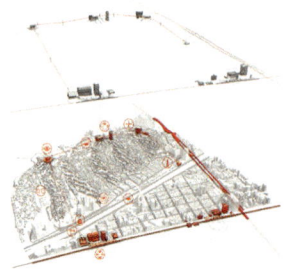

Urban Think Tank _ **Metrocable**
Caracas, 2007 - 2010

Coloco _ **Jardin DeMAIN**
Lemasson Montpellier, 2010

## Concluding remarks: prospects for future inquiries

In the world of architecture and urban design new paradigms are currently change the way professionals, policymakers and citizens thinks about or cope with current global uncertainties, climate urgencies, economic crises, quality of urban life, which affect the way we foresee our future. In a scenario, where the scale and pace of market-driven urbanization and infrastructural development is the only territorial "driver" to be considered, the operative capacity of landscape infrastructures oppose with a position that links human index development and the notion of permanence as a basic planning principle. Likewise, the re-conceptualization of nature, technology and infrastructure permit to read the landscape as an operational platform of possibilities (Castells, 2000) determined by the socio-spatial dimension of networks, both in functional terms (space of flows) and in their physical expressions (space of places). New fields of inquiry emerge from the opportunities offered by the reactivation and reclamation of variable patterns (spectral, textural, topological, etc.) related to landscape infrastructure as generator of urban-rural linkages and self-sustainable development. In this way the urban design and landscape projects, acquires a narrative value that depict contexts for their democratic spatial qualities (decided by many, shared by many, made by many) capable of counteracting, at least in part, the wear and tear that affects the places we inhabit. Wear and tear that not only includes uses, perhaps more easily replaced by others, but above all values and meanings that those same places may convey in post-metropolitan society of 21st century.

## Bibliography

Allen S. (1999) 'Infrastructural urbanism', in: idem, Points + Lines. Diagrams and Projects for the City, Princeton Architectural Press: New York, pp. 46-59.

Augè M. (1992) Non-lieux. Introduction à une anthropologie de la surmodernité, Seuil: Paris.

Batty M. (1983) 'On Systems Theory and Analysis in Urban Planning: An Assessment', In Batty M., Hutchinson B. (eds) Systems Analysis in Urban Policy-Making and Planning. NATO Conf, vol 12, Springer: Boston, pp 423-447.

Bèlanger P. (2016) Landscape as Infrastructure: A Base Primer, Routledge: New York.

Branzi A. (2006) Weak and Diffuse Modernity: The World of Projects at the Beginning of the 21st Cent, Skira: Milan.

Castells M. (2000) The Information Age: Economy, Society and Culture, vol.1 'The Rise of the Network Society', Wiley: Oxford.

Corboz A. (1983) 'Le territoire comme palimpseste' in Diogene, v.121, pp. 14-35.

De Carlo G. (1998) 'L'identità del territorio', in Quaderni di Spazio e Società, n.1, Maggioli: Santarcangelo di Romagna.

Garrau J. (1991) Edge City: Life on the New Frontier, Anchor Books: New York.

Gausa M., et all (2003) The Metapolis Dictionary of Advanced Architecture. Actar: Barcelona.

Grabow S., Heskin A. (1973) 'Foundations for a Radical Concept in Planning', in Journal of the American Institute of Planners, vol. 39, pp. 106-11.

Lydon M., Anthony G. (2015) Tactical Urbanism: Short-term Actions for Long-term Change, Island: Washington, DC.

Mostafavi M., Doherthy G. (eds.)(2011) Ecological Urbanism, Lars Muller Publishers: Baden.

Occhialini E. (2005) 'Research, training, project: ILAUD', in Guccione M., Vittorini A. (eds.) G. De Carlo: le ragioni dell'architettura, Electa: Milan, pp. 212-219.

Schröder J., et all. (2009) Agropolis, Munich.

Schröder J., Carta M (eds.)(2017) Territories: Rural-Urban Strategies, Jovis verlag: Berlin.

Shannon K., Smets M. (2010) The Landscape of Contemporary Infrastructure, NAi Publishers: Rotterdam.

Sijmons, D. (2014) 'Waking up in the Antropocene', in Brugmans G., Strien J. (eds.) Urban by Nature. IABR 2014, NAi publishers: Rotterdam, pp. 12-20.

Strang G.L. (1996) 'Infrastructure as Landscape, Landscape as Infrastructure', in Places, vol.10, p.8.

Tafuri M. (1973) Progetto e utopia. Architettura e sviluppo capitalistico, Editori Laterza: Bari.

Vogt G., Kissling T. (eds)(2020) Mutation and Morphosis: Landscape as Aggregate, Lars Müller: Baden.

Waldheim C. (2016) Landscape as Urbanism: A General Theory, Princeton Architectural Press: New York

Los Angeles River. Author: Bart Jaillet (2017).

# HYBRID INFRASTRUCTURES AND THE COMBINATION OF GREY, GREEN AND BLUE: 4 CASES

*Matilde Pitanti*

**Scenario** There are territories where the relationship between infrastructure and landscape is so close that infrastructure can be explored as a type of landscape and landscape as a type of infrastructure. According to Nijhuis, van der Hoeven and Jauslin (2015) these territories can be defined as flowscapes and the hybridisation of the concepts of landscape and infrastructure can redefine infrastructure beyond the strictly utilitarian definition, allowing at the same time spatial design to gain operative force in territorial transformation processes (Nijhuis, van der Hoeven, Jauslin, 2015). Inside these territories flows and movements are central aspects; mobility, green and water infrastructures are the main actors that contributes to shaping the relationship between natural and human systems (Waldheim, Berger, 2008). Three fields of landscape and infrastructure design are significant to provide objectives for a more complete form of urban landscape architecture: transportation landscape infrastructure, green landscape infrastructure and water landscape infrastructure (Nijhuis, 2013). The first field is identified with the infrastructures that facilitate the various modes of transport, energy and information supply, as well as waste treatment; the second is represented by green infrastructures, intended as a set of interconnection networks of green spaces that maintain and develop the values of natural ecosystems; finally, the third field refers to the management of water resources, coasts and rivers, as well as coastal wetlands and estuaries (Nijhuis, Jauslin, 2015). This contribution intends to explore those infrastructural territories, previously described as flowscapes, analysing four projects realized in the last twenty years in which the presence and the overlap of different fields of urban landscape infrastructures have been used to create and implement the relationships between natural and human systems.

**Infra-scapes visions:** Raffaele Raja explores the theme of industrial architecture, describing clearly the historical relationship between industry and transport infrastructures, and highlighting how, from the beginning, the industry was physically dependent on a very complex series infrastructure and services (Raja, 1983). Places near water courses were first preferred because the river provided low-cost energy and constituted a considerably fast communication and transport route for the time, then, with the availability of coal, the railway network was soon added to communication and transport by river. The railway and the water, together with the industry were the backbone of cities. Tony Garnier's Cité Industrielle is a glaring example of how

the river and the railway were the arterial system of a balanced and coordinated organism; this was assumed as an archetypal value for subsequent industrial and territorial planning (Raja, 1983).
Before the industrial revolution many historical infrastructures hybridized uses and functions, creating interactions with the contexts; just think of Ponte Vecchio in Florence or Ponte di Rialto in Venice (De Francesco, 2017); however, the industrial era triggered a large process of expansion of the infrastructures and, at the same time, a process of mono-functionalization of it.
It is from the idea of modern infrastructure that the promotion of new paradigms and the attempt to study their design potential and the possible relationships between infrastructure and context, activate interactions between different systems (Ferlenga, Biraghi and Albrecht, 2012). Gaetano De Francesco (2017) highlights that today infrastructures no longer have urban expansion as their main objective, but rather the recovery, densification and redevelopment of the existing city. In the McCormick Tribune Campus Center of Chicago, built by OMA between 1997 and 2003, for example, a mixed transport and social infrastructure transversely reconnects a portion of campus previously interrupted by the railway line. Recent projects show indeed the intention of reconnecting parts of the city through public space and landscape, often held together by hybrid and multifunctional infrastructures (De Francesco, 2017).

## Mixed approaches and the combination of green-blue and grey

Traditionally urban water management is based only on grey infrastructures, even if it has numerous environmental and economic consequences (Brears, 2018). Alves Beloqui (2020) highlights how a paradigm shift is needed, which should allow adaptation to long-term climate situation, through the improvement of social well-being, quality of life, infrastructure and the integration of a multi-hazard approach.
Green-Blue Infrastructures involve the use of natural or artificial systems to improve ecosystem services and increase resilience, especially to climate risks (Brears, 2018). These are certainly valid tools to rethink infrastructure systems, however, in already high developed urban contexts the most effective solutions to mitigate flood risks and improve system resilience are mixed approaches, which combine Green-Blue and Grey Infrastructures (Kabisch et al., 2017; Browder et al., 2019). By mixing these two approaches the stability of the grey systems can be combined with the multifunctionality and adaptability of the blue-green systems (Casal-Campos et al., 2015, Alves Beloqui, 2020).

## Four hybrid combinations of grey, green and blue infrastructures

**The Olympic Sculpture Park, the Jardines elevados de Sants, Madrid Rio and Medellin River Parks**

In the last twenty years the operations on various infrastructures have been an opportunity to create complex projects that integrate grey infrastructures with social, green and often blue infrastructures. The presence and the overlap of different fields of urban landscape have been used to create and implement the relationships between natural and human systems. In particular, four interesting cases are here reported and compared: The Olympic Sculpture Park in Washington, the Jardines elevados de Sants and the Madrid Rio in Spain and Medellin River Parks in Colombia; the first four started more or less simultaneously, over a period of three years, while the last is more recent, and is the heir of some of the previous experiences.

The project of the Olympic Sculpture Park, developed between 2001 and 2007, is located on an ex industrial site at the water's edge, owned by the by the Seattle Art Museum. The area was used as an oil storage facility and prior to the construction of the park, more than 120,000 tons of contaminated soil were removed while the remaining contaminated soil was covered with embankments. The project addresses the infrastructural issue on different levels, first of all dealing with the environmental remediation, then, reconstructing a green landscape, studying connections on different levels, integrating pedestrian and vehicular traffic with road and rail. The designers, Weiss/Manfredi Architecture/Landscape/Urbanism, describe it as "a continuous constructed landscape for art, that forms an uninterrupted Z-shaped green platform, [...] and rising over the existing infrastructure to reconnect the urban core to the revitalized waterfront". The project is a complex architecture in which the grey infrastructure of roads and railways, the complex green system and the waterfront are integrated together building a new social and civic value. The pedestrian path scans three different types of landscape and connects the exhibition pavilion on one side and the waterfront on the other. The project has an important environmental impact: the morphology of the ground and the green system work together to collect and purify stormwater, while the connection with the water ecosystem is managed through the construction of an underwater slope, that allows the management of water flows and promotes the proliferation of aquatic flora and fauna. Weiss/Manfredi in this project was able to integrate blue and green infrastructure on the grey one, without falling into naive attempts at camouflage, but having the ability to maintain an infrastructural language and creating urban identity, at the same time.

In the same period between 2002 and 2016, the Jardines elevados de Sants have been developed by the Ayntamento de Barcelona and entrusted to the designers Sergi Godia, Ana Molino together with architects Esteyco Ingenieria. It extends for a length of 800 meters and has an average width of 30 meters. The gardens of the Rambla de Sants, as they are also known, were created with the aim of stitching up a piece of the city cut by a railway corridor. The installation of the high-speed railway line (AVE) was an opportunity to bury two tracks, necessary for the new AVE infrastructure (Passatges bcn, 2014). The architects Sergi Godia and Ana Molino from the beginning renounced the possibility of burying the entire railway section due to technical and economic problems. Rather, they imagine a long and elevated walk, similar to New York's High Line, which over time can be extended to

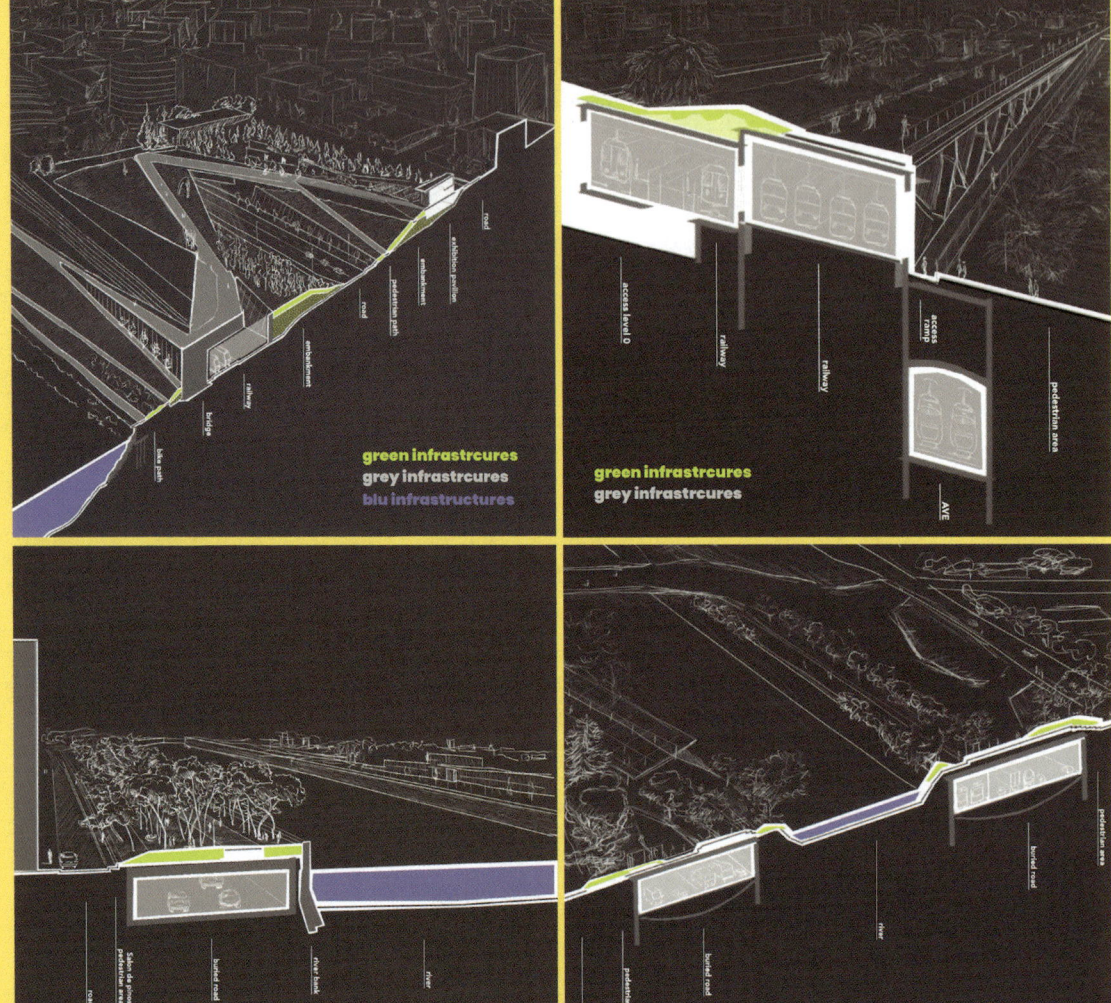

the nearby cities of Hospitalet, Esplugues and Cornellá, to give shape to a 5 km long promenade plantée (De Francesco, 2017b). The railway infrastructure is incorporated within a prefabricated concrete and glass structure, characterized by diagonal beams, in the shape of Warren beams, which speak a typical language of the infrastructure, allowing at the same time visual permeability and reducing noise pollution. The roof is a panoramic garden, consisting of two linear paths characterized by different vegetation (De Francesco, 2017b). The civic infrastructure certainly represents an improvement in the degraded urban context, however, probably due to the lack of a consultative process and shields for houses, at the beginning the project caused protests and the loss of value of some properties. Unlike the High Line, in the Jardines elevados de Sants, the railway is not abandoned, and compared with the next cases of Madrid and Medellin, the grey infrastructure is not buried, but it is integrated into the new urban design. It involves the reconversion of an entire district, which is not limited to the line of the tracks, but, thanks to the availability of important pieces of public land on the side, it expands and integrates into the city, and connecting with the others public green spaces, aims to create a new urban green corridor.

A recurring operation in the last twenty years was burying high-speed roads in urban contexts, to restore areas to public spaces and to create valuable real estate and enlarged public realms (Shannon, Smets, 2010). One of the most famous river-sides urban redevelopment of recent years is certainly Madrid Rio and the recovery of the waterfront on the Rio Manzanares. Between 2003 and 2007, the city of Madrid buried a section of M-30 road, built parallel to the Manzanares river, regaining a large area along the river, that has been the subject of a design competition in 2005. The competition was won by the M-Rìo team, that brought together Burgos & Garrido studio, Porras & Lacasta, Rubio & Alvarez Sala and West8. In this case the coverage of the road was the opportunity to create a more complex infrastructure package, which contained the road on the lower level and a social and green infrastructure on the upper level. The project, completed in 2011, creates a series of green spaces along the Manzanares, reconnecting the city with the river and sewing together the northern and south-eastern parts of Madrid, sacrificed due to the motorway. The mixed infrastructure connects the city not only in a longitudinal sense but also in a transversal sense, integrating a large system of crossings and bridges within it . This large hybrid project crosses the whole city with a completely public space, a transport and civic infrastructure, in which various services are located, including areas equipped for sports activities, skate parks and play areas for children.
Medellin River Parks projects looks directly to a series of other operations such as Madrid Rio. It is, in fact, a large urban transformation project carried out after the decision of burying high-speed roads inside the city and following a design competition. The Colombian city develops within a river valley, crossed by the

Medellin river. Once the watercourse was artificially channelled, the valley underwent a strong infrastructural development and a heavy urbanization. Over the years the river has become an important infrastructural axis, flanked by roads and highways on both sides, which divided the city physically, socially in the environment. In 2011 a new urban plan has been adopted and Medellín River Parks was then created (Sáenz, 2016). The operation, that started in 2014 and it is on its second phase in 2020, recently won first place at the II Latin American Biennial of Landscape Architecture.

The project was developed by Sebastián Monsalve, Juan David Hoyos, together with Nicolás Hermelín, Luis Vera, in a team made up of architects, landscape architects and engineers. It aimed to integrate engineering, urban planning and landscape, to create a recompositing of the urban, environmental and social integration. The complex urban project covers the roads with an environmental and social infrastructure; a series of urban parks and two pedestrian bridges give the inhabitants the opportunity to get closer to the river. River Parks was an opportunity to create a hybrid project between grey and green infrastructure, to restore quality of habitability to abandoned spaces on the banks of the river.

## Final considerations

Stan Allen (2014), quoting Jim Corner, highlights how urban infrastructures are the basis of future possibilities, a preparation of surfaces for future appropriation in contrast with the merely formal interest of the single object. The urban infrastructure project, he writes, is a strategic project and not just a compositional one. In all four projects analysed the transport infrastructure is not dismissed, but it is an active part of the urban system, becoming no more factor and cause of degradation, but an opportunity for urban redevelopment and reactivation. The presented projects, even if with some geographical, technical and dimensional differences, are all hybrids, which integrate the grey, green and/or blue components into a single social and civic infrastructure, which is the axis and driver of a new urban identity.

*What is required is a new mindset that might see the design of infrastructure not as simply performing to minimum engineering standards, but as capable of triggering complex and unpredictable urban effects in excess of its designed capacity. That is to say, infrastructure creates concentrations of density that in turn trigger concentrations of activity.*
— S. Allen, 2014

# Bibliography

Allen, S. (2014), "Landscape Infrastructure", in Area, vol. 127, Identity of the landscape, pp. 10-11.

Allen, S. (1999), "Infrastructural urbanism", in: idem, Points + Lines. Diagrams and Projects for the City. Princeton Architectural Press, New York, pp. 46-59.

Alves, A. (2020), Combining green-blue-grey infrastructure for flood mitigation and enhancement of co-benefits. CRC Press, Balkema.

Brears. R. C. (2018), Blue and Green Cities. The Role of Blue-Green Infrastructure in Managing Urban Water Resources. Macmillan Publishers, London.

Browder G., Ozment S., Rehberger Bescos I., Gartner T., Lange G. M. (2019), Integrating green and gray: Creating Next Generation Infrastructure. World Bank and World Resources institute, Washington DC.

Casal-Campos A., Fu G., Butler D., Moore A. (2015), "An Integrated Environmental Assessment of Green and Gray Infrastructure Strategies for Robust Decision Making", in Environmental Science and Technology. vol 49, 14.

De Francesco, G. (2017a), Infrastrutture dell'acqua. Strategie adattive all'emergenza idrica dei mutamenti climatici. Progettare infrastrutture idriche di nuova generazione. PhD. Thesis.

De Francesco, G. (2017b), "I jardines elevados de Sants a Barcelona. Un'infrastruttura contemporanea", in L'industria delle costruzioni, vol. 454, pp. 98-103.

Kabisch N, Korn H, Stadler J, Bonn A. (2017), Nature-based Solutions to Climate Change Adaptation in Urban Areas. Springer.

Nicolin P. (2009) "Olympic sculpture park", in Lotus international, vol. 139, Editoriale Lotus, Milano.

Nijhuis, S., & Jauslin, D. (2015). "Urban landscape infrastructures. Designing operative landscape structures for the built environment", in Research in Urbanism Series, vol. 3, issue 01, p.13-34.

Shannon, K., Smets, M. (2010), The landscape of contemporary infrastructure, nai Publishers, Rotterdam.

Nijhuis S., Van der Hoeven F., Jauslin D. eds. (2015). "Flowscapes. Designing infrastructure as landscape", in Research in urbanism series, vol. 3, TU Delft, Delft, the Netherlands.

Nijhuis, S. (2013) "Principles of landscape architecture", in Farina, E. & S. Nijhuis (eds.) Flowscapes. Exploring landscape infrastructures. Mairea Libros Publishers, Madrid, pp. 52-61.

Raja, R., (1983), Architettura industriale: storia, significato e progetto. Edizioni Dedalo, Bari.

Waldheim, C., Berger A. (2008) Logistics Landscape, in Landscape Journal, vol 27, issue 02, pp. 219-246.

# On-line sources

El Cajón de Sants in Passatges bcn, [Online] 12 August 2014. [Accessed 24.04.20] Available at: http://passatgesbcn.blogspot.com/2014/08/el-cajon-de-sants.html.

Jardines elevados de Sants en Barcelona / Sergi Godia + Ana Molino architects, in Plataforma Arquitectura. [Online] 15 dic 2016. [Accessed 24.04.20] Available at: www.plataformaarquitectura.cl/cl/801124/jardines-elevados-de-sants-en-barcelona-sergi-godia-plus-ana-molino-architects.

Jardines elevados de Sants, Barcelona, in Area-arch, [Online] 22 November 2016 [Accessed 24.04.20] Available at: www.area-arch.it/jardines-elevados-de-sants-barcellona/.

Lopez, H. (2016) La invasión de la intimidad del cajón ajardinado de Sants llega al juez, in El periodico [Online] 8 November 2016. [Accessed 23.04.20] Available at: www.elperiodico.com/es/barcelona/20161107/un-vecino-del-cajon-de-sants-demanda-al-ayuntamiento-por-falta-de-intimidad-5613157.

Mele F. (2014) Da autostrada a parco sul fiume: il caso felice di Madrid Rio, in Artwork [Online] [Accessed: 25.04.20]. Available at: www.artwort.com/2014/04/11/architettura/autostrada-parco-fiume-caso-felice-madrid-rio/.

Minner. K. (2011). Olympic Sculpture Park / Weiss Manfredi, in ArchDaily [Online] 06 Jan 2011 [Accessed 25.04.20], Available at: www.archdaily.com/101836/olympic-sculpture-park-weissmanfredi.

Montilla, R. (2020), El Ayuntamiento de Barcelona, condenado a pagar por el cajón de Sants, in La Vanguardia. [Online] 6 July 2020. [Accessed 7.07.20] Available at: www.lavanguardia.com/local/barcelona/20200706/482138165450/ayuntamiento-barcelona-pagar-cajon-sants.html

Ortega, S. (2015), Chaos? This is open-heart surgery: Medellín risks a massively expensive plan to bury its highway, in The Guardian [Online] [Accessed:15/02/2020].

Sáenz. L. (2016), ¿En qué está el proyecto Parques del Río en Medellín? ArchDaily Colombia, [Online] 19 October 2016 [Accessed: 15/04/2020], Available at: www.archdaily.co/co/797527/en-que-esta-el-proyecto-parques-del-rio-en-medellin

Un cuore verde per Seattle, in Domusweb [Online] 23 May 2012 [Accessed 24.04.20], Available at: www.domusweb.it/it/architettura/2002/05/23/un-cuore-verde-per-seattle-.html.

Madonna of Chancellor Rolin (excerpt), Jan Van Eyck (1453). Musée du Louvre, Paris.

# LANDSCAPE INFRASTRUCTURES CROSSOVERS: OUTLINES FOR AN ANNOTATED BIBLIOGRAPHY

*Beatrice Moretti*

«È che per questo paesaggio, moderna Terra di Babele, l'intercambiabile, l'indecifrabile, l'inconoscibile, lo sterminato, sembrano le sole definizioni possibili.»[1]

- L. Ghirri, Per un'idea di paesaggio, 1986

1. "It is that for this landscape, modern Land of Babel, the interchangeable, the indecipherable, the unknowable, the endless, seem to be the only possible definitions." (Translation by B. Moretti). Cf. Ghirri L. (2021)[1986], *Niente di antico sotto il sole. Scritti e interviste.* Macerata: Quodlibet, p. 137-138

Before landscape was a specialist profession, an academic discipline, a medium for design, it was a subject for the theatre arts, a mode of human subjectivity and a genre of painting. Ubiquity, paradox and promiscuity are some of the characters that, even today, can be attributed to the concept of landscape.
Its epistemic potential – as well as its applicative one – has not yet been exhausted and motivates the multiplication of interdisciplinary theories about it. Those interwoven here aim to highlight its evolution and relationship with the concept of infrastructure.
Not claiming to be exhaustive, they construct a compact set of references – an annotated bibliography – referring to the decades between the 20th century and the 2000s.

## Landscape as a Phenomenon

2. Cf. Jakob M. (2005). *Paesaggio e Letteratura.* Firenze: Leo S. Olschki

3. Cf. Jakob M. (2009). *Il Paesaggio.* Bologna: Il Mulino

Landscape painting emerged during the Renaissance in the Nordic Flemish countries, laying the foundations for the modern meaning of landscape in figurative and literary arts. It is precisely in this historical moment, in fact, that landscape architect Michael Jakob places the emergence of a social and cultural landscape conscience, linking its origin to an *experience*.
Even if it is not possible to identify one single reason for the emergence of a landscape conscience, it is true that already in very ancient times (in the Hellenistic period when the structure *polis – agros* was first defined[2]), the civilized human being – in Jacob's words, the citizen – *experienced* a substantial broadening of the horizons of his geographical knowledge and, thus, came into contact with what is considered *the other*, i.e. the nature. In order to become aware of it, the modern citizen develops a new *way of seeing*, through which to project personal desires and meanings. Regardless of the countries and speeds at which it developed, then, the landscape conscience has a historical dimension which enables a proto-history of the landscape to be traced.[3]
It is through two famous paintings, realized between the 15th and 16th centuries, that the image of the landscape (the "landscape-image" now widely flaunted, according to Jakob) begins to assert itself and define its fundamentals. In the *Madonna of Chancellor Rolin* (1435), the Flemish painter Jan van Eyck depicts nature as an object of desire: to do so, he contrasts two points of view and constructs a

sophisticated device for observation through architecture.
The interior of a sumptuous palace – a closed and controlled system – houses the point of view of theology embodied by the figure of the Chancellor, the Virgin and of several allegorical signs intended to reinforce the absolute primacy of religion. In the background, the garden-terrace opens up a second viewpoint towards nature: it is in fact the two tiny and unnamed figures, depicted from behind, that represent van Eyck's pictorial reflection and hold up the whole composition. Through their eyes – pointed outwards, towards open spaces – the relationship between observer and nature is defined. While in van Eyck's painting, the landscape is filtered by the observers' gaze, in Giorgione's *Tempest* (1506-1508) the human figure takes on a decisive role in the reading of nature. Although no longer at the center of the composition, the figure occupies a symbolic interface space in the artistic representation: it does not direct the spectator's gaze but turns towards the spectator in an unexpected *face-to-face*. The meaning of the landscape is hence revealed through the messages sent by the figures on the scene: in Giorgione's work – the first to be identified as "di paese" (of the village), that is "di paesaggio" (of the landscape) – the woman with child on the right breaks through the *fourth wall* of the composition and allows a direct identification thanks to anyone can *enter* the landscape and understand its purpose.

Using this historical reading, Jakob describes landscape as a phenomenon. Not measurable or existing in itself, landscape is the product of a set of actions on which its perception depends. It is an artificial phenomenon as it is mediated by a complex series of relationships. It is also subjective because represents the outcome of experiences perceived by subjects who act as the constituent basis of the phenomenon itself, becoming observers.
Landscape, finally, is not an analogue of art (but almost the opposite): it is in fact through landscape texts and paintings that one can have an anticipatory and catalytic experience of *real* landscapes and transfer ideas, values and models onto them.

Already in Jakob's idea of landscape as a visual cutout made up of the gaze of social subjects, the power of vision emerges as a central tool for a tentative definition of the phenomenon. It is, after all, the era of the dominance of perspective: the effects of distance, depth and vanishing point open up endless horizons that expand figurative possibilities and quests for knowledge. Geographer Franco Farinelli argues that, until Ptolemy, geographical representation – i.e. the reduction of the world to a map – concerned only the things that *could be seen*.[4]
The act of seeing coincided with the appropriation of a space, attributing profound importance to the knowledge of vision.
However, for many centuries, the syntax of the modern territory did not appear on geographic drawings: land routes and infrastructures were not clearly represented or, more often, were assimilated to the paths of natural watercourses. This is because in the Middle Ages and later on, maps were *a copy of the world* and reflected the

**Landscape or the World in the form of a Place**

4. Cf. Farinelli F. (2003). *Geografia. Un'introduzione ai modelli del mondo.* Torino: Piccola Biblioteca Einaudi; Farinelli F. (2009). *La crisi della ragione cartografica.* Torino: Einaudi

relationships of which the world was composed. It is only in the 20th century that, with the advent of modern culture, this situation is totally reversed and the world is finally reduced to an image, i.e. to a map; as Farinelli acutely argues «this is how the world is really transformed into the face of the Earth» (2003: 15).

In this age of *the image of the world*, landscape is for Farinelli a way of seeing and looking at the things of the world or, even better, is the way in which modernity sees the world in the form of a place, through a representation that obeys a relationship of an iconic nature.

In order for a landscape to exist, then, the elements previously identified through painting (the subject that looks and the object to be looked at) are not enough but it is essential to have the greatest possible horizon, ideally a height that functions as a vantage point. The supremacy of vision, however, is not destined to last. «With the advent of miniaturization, dematerialization and computerization [...], a world is being produced in which, for the first time, the domain of vision returns almost nothing meaningful about the mechanisms and activities that govern its operation» (2003: 53). Thus, the relationships between functioning systems, infrastructure, environment and human beings become virtual and latent: the representation and, consequently, the capacity to decipher the world around us enter into crisis. The need for other types of landscapes – not exclusively related to what we can experience with our eyes – increases exponentially, pushing urban studies to renew their terminology and design tools.

## Landscape as a Model for Urbanism

The discourse about the concept of landscape in urban studies went through a fundamental turning point at the end of the last century when, especially in the North American socio-economic context, the theme of the city began to be approached through the lens, or multiple lenses, of landscape (at that time, in fact, many professional landscape firms were beginning to win major commissions). In the American academy of the early 2000s, this was also a disciplinary issue to position landscape architecture in relation to urbanism and planning. In an attempt to address this questions, landscaper Charles Waldheim edited two emblematic volumes[5] through which he traced the features of the *Landscape Urbanism*'s movement. «Landscape has been claimed as a model for contemporary urbanism. » (2016: 13). With these words, Waldheim declares the disciplinary realignment to which the theories of *Landscape Urbanism* aspire: an implicit criticism of architecture and urban design whose historical role in determining urban form is supplanted. Waldheim draws on the contributions of authors who have discussed the newly relevant capacity of landscapes to organize horizontal surfaces – i.e. to produce urban effects traditionally achieved through the construction of buildings – not only in terms of configuration but specifically in terms of materiality and performance. With the involvement of disciplines traditionally distant from city making (literature, painting, photography, ecology, etc.),[6] landscape becomes a medium uniquely able to respond to temporal changes, environmental disasters and adaptations. There

5. Cf. Waldheim C. (2006) (ed.). *The Landscape Urbanism Reader*. Princeton and Oxford: Princeton Architectural Press; Waldheim C. (2016)(ed.). *Landscape as Urbanism*. Princeton and Oxford: Princeton Architectural Press

6. Cf. Doherty G., Waldheim C. (eds.)(2015). *Is Landscape...? Essays on the Identity of Landscape*, Routledge, Oxford

Katy Freeway, Houston, TX, USA, 2015. The visible section stretches ca. 10 km from Silber Rd to Sam Houston Tollway.

Newark International Airport and Port Newark–Elizabeth Marine Terminal, Newark, NJ, USA.

Deltaworks, the Eastern Scheldt storm surge barrier under construction, 1954-1997, NED.

 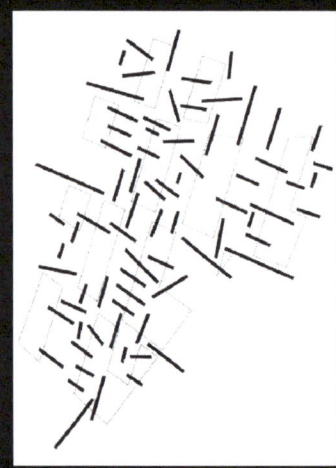 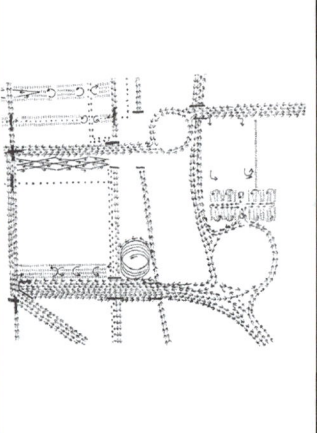

Field conditions diagrams by Stan Allen. Source: Points + Lines. Diagrams and projects for the city, 1999.

Logistical Activities Zone (ZAL), Barcelona by Stan Allen, 1996. Source: Points + Lines. Diagrams and projects for the city, 1999.

Philadelphia Planning Study: movement diagrams (excerpt) by Louis Kahn. Source: Points + Lines. Diagrams and projects for the city, 1999.

are many design examples to which Waldheim refers in his studies, demonstrating the existence of ongoing processes that need to be codified. One is definitely Barcelona where the focus of design actions, which in the 1980s and 1990s transformed the traditional city center, shifted in the early 2000s to the airport, the logistics zones and the industrial and metropolitan waterfront and river ways. Such large-scale infrastructural landscapes reveal their decisive role as elements of the urban infrastructure, confirming how *Landscape Urbanism*'s theory triggers a major revolution in the way we think about highly infrastructured areas. This theory, in fact, contributes to stress issues such as the forms of landscape generated by the complex set of traffic and logistic flows that support and serve territories and cities: new spatial conformations that have been defined over the years through concepts like *operative* or *distribution landscapes*.[7] Waldheim contributed to these reflections by naming them *logistics landscapes*, i.e. artificial territories characterized by new industrial forms structured around global trade networks and vast areas destined to accommodate the import/export process and the transport and delivery of goods. Talking about landscapes and infrastructural complexes, Waldheim maintains that «by describing this logistics landscape in spatial and economic terms, it may be possible to apprehend the forms that it takes, to anticipate the priorities that it pursues, […] and to acknowledge our embeddedness in the culture it represents» (2003: 53).

7. Cf. Nigra Snyder S., Wall A. (1998). "Emerging Landscapes of Movement and Logistics", *Architectural Design Profile*, no. 134, pp. 16-21

## A Shift towards Infrastructure

A paradigm shift in the recognition of contemporary urbanism was certainly signaled by the first-and second-prize entries of the "Urban Park for the 21st Century" competition for Parc de la Villette (Paris, 1982). Both the winning proposal of Bernard Tschumi's office and that of Rem Koolhaas and OMA were significant in the development of landscape urbanism as design strategy, especially to deal with evolving arrangements or postindustrial sites. As is now well known, the competition for la Villette began a path of large-scale projects in which landscape was conceived as a complex medium through which articulate urban relationships of an infrastructural nature; as Stan Allen has argued, however, in these cases the landscape did not constitute a formal model for urbanism, but rather a model for *process*.[8]

Following these ideas, Allen highlights the lack of connection, at least in postmodernist American architecture, between the work of architects and the construction of urban infrastructure: in his beliefs, architects have retreated from instrumental questions related to technique, implementation and, generally, from the questions of infrastructure. Hence, the notion of infrastructural urbanism offers a new sense of architecture's potential to plan the future of the city by building a new model for practice: not surprisingly, Allen states that «infrastructural urbanism understands architecture as *material practice*.» (1999: 52)

Issues such as mobility, territorial organization, local ecologies have always been part of traditional architectural skills (before

8. Cf. Allen S. (2001). "Mat Urbanism: The Thick 2-D", in Sarkis H. (ed.), *CASE: Le Corbusier's Venice Hospital*, Munich: Preste

The Tempest, Giorgione, (1506–1508) Gallerie dell'Accademia, Venice.

the separation of planning disciplines) and, over the years, have been employed to deal with problems on a large scale. These specialties can be demanded in the field of architecture and implemented with the new technologies available. Allen proposes the idea of strategically making use of the typical characteristics of infrastructures (detailed design, standard elements, structural repetitiveness, regularity, etc.), thus facilitating an *architectural approach to urbanism*. Infrastructures have a fundamental role in the constituting of functioning systems that can establish relationships and, at the same time, modify the spatial component. They possess flexibility, forecasting and adaptability, qualities that are transferred to spaces in a search for a sort of indefiniteness that, given to the structures, can ease future technological adaptations.

## Landscape as Infrastructure

9. Cf. Bélanger P. (2017). *Landscape as Infrastructure*. London: Routledge

Through Pierre Bélanger's investigation,[9] the relationship between landscape and infrastructure at the turn of the last century becomes more powerful. By acknowledging the need for a more up-to-date and contemporary understanding of infrastructure, Bélanger proposes to expand it as a landscape of systems, services, scales, resources, flows, processes that sustain and cultivate urban economies. To do this, it is essential to implement a landscape practice and conceive it as an integrative and *horizontal discipline*. According to historian Rosalind Williams, one literally needs to set up the "landscape" vision (i.e. the horizontal one) in order to include in landscape's sphere of intervention the operational, logistical and infrastructural aspects of today's urbanization. Undoubtedly, the drive towards an ecological rethinking of these highly-engineered environments depends on massive and global transitions from industrialization to urbanization and, even more, on the rise of environmental concerns since the 1970s and the crisis of public works planning in the 1980s. In this line of reasoning, Bélanger raises the following provocative questions: «is landscape infrastructure? That is, can we consider the "non-mechanical," "non-linear" and "non-stable" media of living systems as infrastructure? And conversely, is infrastructure landscape? That is, can we consider the non-biologic, non-dynamic, and non-adaptive material of infrastructure as a constructed landscape and lived experience?» (2015: 193).

The set of new design strategies and models of practice triggered by the conjecture of *landscape as infrastructure* (and its reverse, *infrastructure as landscape*) reveal the potential of infrastructure as a great enabler. Able to function as both a *response to* and a *generator of* horizontal forms of development, the infrastructure is the interface through which it is possible to interact with the biological and technological world. Conceptualised through the filter of infrastructure, landscape is reformulated as an instrumental system of essential resources, services and agents that generate and support urban dynamics. Moreover, in order to achieve the pressing demands of the energy transition and the UN's 2030 Agenda for Sustainable Development (17 Sustainable Development Goals - SDGs, 2015), landscape becomes sophisticated, specialized and new performances are required. In the light of this, the bond – almost a conceptual and strategic overlap – between landscape and infrastructure appears crucial to build an increasingly integrated and interdisciplinary design for man-made spaces.

> «This double entendre of landscape infrastructure entails both the design and un-design of urbanization through new faculties and facilities. [...] Rather than propose a universal theory of new ideology here, these reciprocal possibilities provide simple and sometimes subversive practices that support new attitudes and appetites for crossovers, by design, by improvisation, by coincidence, and by accident.»
>
> - P. Bélanger, *Systems of Systems*, 2017

# THE CHERNOBYL EXCLUSION ZONE.
# OR INFRASTRUCTURE AS MEMORY

*Davide Servente*

On the night of April 26, 1986, in northern Ukraine, reactor number 4 of the Vladimir Ilyich Lenin Nuclear Power Plant in Chernobyl explodes during a 'safety' test. The release of 50% of the radioiodines and 30% of the radiocesium into the atmosphere causes a concentration of radioactivity in the environment between 50 and 250 million Curies, an amount approximately 100 times greater than that of the bombs dropped in 1945 on Hiroshima and Nagasaki. Following the declaration of a state of emergency, 350,000 people residing in the areas close to the explosion are evacuated, and an area of 2,600 hectares (equal to 3,641 soccer fields) between Ukraine and Belarus is declared to be the *Зона отчуждения Чернобыльской АЭС* (Chernobyl Exclusion Zone) in which inhabitation, agriculture, hunting, and industry are prohibited. Located less than a kilometer from Chernobyl and built to house construction workers and plant employees, Pripyat was designed as an exemplary socialist city. Defined as a large pan-Soviet project, the city was also known by the name of Atomograd, an example of utopian urban planning inspired by modernist principles. The city was characterized by tall residential buildings alternating with wide, tree-lined avenues, with state-of-the-art public services in terms of design and construction techniques.
Pripyat is still completely uninhabited by humans. Once characterized by its numerous green spaces and well-tended gardens, Pripyat has now been 'attacked' by vegetation. Overgrown, the trees have overtaken all of the open space, and a dense moss covers the ground floors of the buildings and the streets. Having remained unchanged since being abandoned by its inhabitants and populated only by wild animals that move in total freedom, Pripyat could today be designated the greenest city in the world, a deserted 'anti-Garden City' in which the vegetation is inhospitable and harmful to one's health. The buildings in ruin, the absence of people, and the atmosphere of decay make Pripyat the "dead city *par excellence*" (Belpoliti, 2005: 77). Yet, if in our cities, abandonment is a suspensive condition that corresponds to an expectation of rebirth, in Pripyat, death is everlasting: "The ruin continues to ruin incessantly, in a form both settled and unsettled, as if the death throes of the city itself were something even stronger than its own definitive death." (Ibid.)
In March 2002, the Ukrainian team of developers for GSC Game World entered the Chernobyl Exclusion Zone to document the landscape surrounding the plant's ruins and to reproduce it digitally in the first-person shooter video game *S.T.A.L.K.E.R.: Shadow of Chernobyl* (2007). About nine years later, the soldiers guarding the perimeter of the 'zone' chased the *S.T.A.L.K.E.R.* fanboys for several nights, who were armed with air rifles and equipped with conspicuous gas masks,

Pripyat, the day of the evacuation. 27 April 1986. Sergei Yakunin Archive. Source: *pripyat-city.ru*

intent on reproducing the dynamics of the game in the real places that were its inspiration. The video game has made Chernobyl an "augmented landscape" in which virtual shots overlap with a perennial silence, and pixels replace radioactive particles (Young, 2016: 132). Pripyat oscillates between "two cities superimposed like an urban moiré" (Ibid.), a simultaneously virtual and material place, a digital memory and a ruin. The ruins of the Chernobyl nuclear plant are the memory of a future never reached, in which progress is not the bearer of innovation but instead represents the nostalgia of an uncorrupted past in which the present is suspended in a hyperreal fixed state.
In 2002, the Canadian-American photographer Robert Polidori was one of the first to access the 'zone' and to document its state of abandonment. His photos compiled in the book *Chernobyl, Zones of Exclusion* (2003) contributed to creating the new image of Pripyat, depicting the more emotionally engaging places – kindergartens, schools and the amusement park that would have been inaugurated on May 1, 1986 – and the abandoned everyday objects.
Today, tourism is the only economic activity in the area: with the reduction of radiation since 2011, it is possible to visit the principal sites affected by the disaster accompanied by a guide at a cost of 160 euros. In 2019, Chernobyl and Pripyat saw a record number of 100,000 visits. The attention paid to the Exclusion Zone is also evidenced by the 65.900 followers of the Instagram profile *@chernobyltour* – one of the agencies that organizes visits to the remains of the plant – equal to those of *@veneziaunica*, the profile of the tourism office for the Italian city of Venice [updated March, 2022].
The ruins are the only human artifact that, although deteriorated or abandoned, offer an aesthetic experience. Some of them obtain a

certain level of fascination and power of attraction because what produced them has more important historical relevance than the ruins themselves (Whitehouse, 2019: 16-18). This romantic attraction today finds a new and amplified significance in social networks. Abandoned sites, settings of conflict or objects of disaster are the destination of a 'dark' kind of tourism that goes beyond research or personal growth. "Ruin porn," "disaster porn," and "war porn" are the terms used to define the practice of photographing and sharing images of places that have been the scene of tragedies or conflicts (Ibid., p. 11), like the photos of Dresden in the aftermath of the bombing that razed it to the ground at the end of the Second World War or those of the crater created by the collapse of the Twin Towers. To these can be added abandoned infrastructure and industrial buildings that, due to their structural form, fail to find a new role, except that of bearing witness to the technological conquests – and the defeats – of mankind.

The ruins of Chernobyl and Pripyat, like the Imperial Forum of ancient Rome or the remains of Machu Picchu, bear witness to an extinct society. What makes this revolution clear is the silent presence of architecture and its laid-bare resistance to the passage of time: abandonment and ruin as symbols of "an absence that can never be overcome" (Huyssen, 2003: 69). In Chernobyl, like in Pripyat, everything is ruin, "petrified remainders of an overwhelming trauma that challenge any desire to forget" (Dobraszczyk, 2017: 267).

Fifteen days after the explosion, the fire extinguished, construction began on a steel and concrete structure – called a sarcophagus – to 'entomb' the destroyed reactor and to prevent the radioactive material still present from escaping. Completed in 206 days, the imposing structure with a crude shape and high buttresses like a Gothic cathedral is "a crypt meant to cover up the burial chamber of

Left: Chernobyl before the accident. Source: pripyat-city.ru
Right: Power-generating Unit 4 of the Chernobyl Nuclear Power Plant after the explosion.©The Museum of Russian Art.

the atomic era" (Virilio, 2006: 162), an involuntary monument to the competition between man and nature in which man is the inevitable loser. In 2019, after four years of work, a second giant structure is completed to cover the sarcophagus and permanently seal in the exploded reactor. At 275 meters long, 108 meters wide, and 92.5 meters high, the enormous vault is the first useful act in the definitive dismantling of the reactor and the removal of the remaining waste.
In July 2016, the Ukrainian Ministry of Ecology announces its intention to redevelop the area affected by the explosion – within a radius of 30 kilometers from the plant – by converting it to renewable energy production. At the initiative of the Ukrainian-German company Solar Chernobyl, at a distance of around a hundred meters from the sarcophagus, the one-megawatt photovoltaic park costing €1,000,000 is now being completed. The new facility will utilize the high-voltage transmission lines once used to distribute the electricity produced by the plant. The Ukrainian government has also made 25 square kilometers of land available for further rehabilitation projects. Articulated in the "green" sense of renewable energy, sustainability today appears to be the only solution for revamping the image of the area and perhaps bringing humans back to inhabit the disaster area one day.
Assuming that the Exclusion Zone once again returns to being inhabited, the buildings in Pripyat could have two possible fates. If considered as a memory of a vanished society, they should be kept as archaeological finds and musealized like the cities of Pompeii and Herculaneum. The abandoned objects and furniture would be collected, catalogued, and stored in new buildings containing services for researchers and tourists. Itineraries for visits would

be created, equipped with infographics, and cafés and bookshops opened. Or Pripyat could go back to being a model city, but brought up to date based on the current debate over ecological concerns that permeates urbanism and contemporary architecture. Taking advantage of the luxuriant vegetation, the buildings would be completely covered with trees, contributing to the green policy already underway, improving the homes' climate control, saving water, and increasing the already thriving biodiversity of the area: a large park such as the Landschaftpark in Duisburg but without the costs of re-naturalization.

In both scenarios, ordinary buildings and infrastructure, devoid of architectural quality, would be saved and passed on to future generations. Conservation that would not only involve infrastructure and architectural artifacts – lending them an aesthetic quality and a cultural value that were absent at the time of their creation – but above all the contemporary landscape, the nature that has re-appropriated the space once modified by man, and which has brought life back where the failure of technology had denied it.

A Pripyat museum in itself would favor its status as a ruin with historical and cultural value, relaunching it as a tourist destination. Pripyat as a "green city" would reverse its present situation, making it a paradigmatic case of urban redevelopment through the most extreme real estate operation in history.

Pripyat is a case on the outer margins, but not the only one. In the Fukushima exclusion zone following the nuclear disaster in 2011, ten cities were evacuated and 184,670 people removed from their homes. Like in Chernobyl, vegetation has conquered all the open space, and wild animals roam undisturbed among the buildings. Inaugurated even if not yet finished in 1982 in Juragua, Cuba, Ciudad Nuclear was intended to be a model socialist city – like Pripyat – and to house workers and scientists working on the nearby nuclear power plant under construction. Once the funding by the Soviet Union ceased, the facility was never completed, and the city that remained unfinished is now partially inhabited. There are also countless ghost towns, large-scale ruins produced by climate change, economic crises, or conflicts, such as the tourist district of Varosha in the city of Famagusta in Cyprus, uninhabited since 1974 due to the Turkish invasion. Or the countless residential neighborhoods and industrial facilities in Detroit, abandoned in the early 2000s after the city's economic collapse. Or the Chinese village of Houtouwan on the island of Shengshan, abandoned since the 1990s due to the difficulty of connections with nearby Shanghai and completely covered by vegetation, which has become a tourist destination.

Ruins, whether old or new, spectacular or ordinary, hidden or visible, offer themselves as the foundations of new planned cities. Ordinary infrastructure and buildings that acquire a new meaning after their abandonment.

The resistance to the passage of time has made the ordinary buildings of Pripyat a tangible memory of the disaster that overwhelmed them, of a catastrophic event that marked the end of humanity's unconditional trust in the promises of its technological

Chernobyl today. Source: Alamy.

progress. Yet, their ordinariness conveys an image in which one can recognize oneself and be recognized, characteristic of everyday life and reassuring because it is a part of everyone's experience, because it tells a story that has already been heard, and it is understood as bearing witness to life based on collective memory. Reclaiming these places implies the attribution of a new cultural meaning to artifacts without a material value, recognizing their role as the memory of a recent past and as the foundations for a near future.

Attempting to envision a future for the Exclusion Zone is meant to be an opportunity to reflect on the reuse of ordinary buildings and abandoned infrastructural heritage; it means thinking about a third way that goes beyond the mere conservation of artifacts with a recognized historical, artistic, or technological value and that also includes ordinary buildings that belong to our daily environment. That is, recognizing their value, for local and non-local communities, as "intrinsic subjects of architectural and urban imagination" (Will: 2020, 15) and to moderate "the threatening dynamics of growth towards the proclaimed balanced state of a constantly regenerating and self-sustaining system" (ibid.).

But today, life in Pripyat is still suspended, only imagined or remembered by those who left it, on April 27, thirty-five years ago, convinced that they could return there after a short time. Together with their belongings, the present of these individuals has remained in a city where time does not flow.

*What lingers most in my memory of Chernobyl is life afterwards: the possessions without owners, the landscapes without people. The roads going nowhere, the cables leading nowhere. You find yourself wondering just what this is: the past or the future.*

- S. Alexievich, 2016

## Bibliography

Alexievich, S. [1997](2016). *Chernobyl Prayer: A Chronicle of the Future*. London: Penguin Classics.

Belpoliti, M. (2005). *Sarcophagus*. Domus, 884. p. 76-77.

Brown, K. (2013). *Plutopia. Nuclear Families: Atomic Cities, and the Great Soviet and American Plutonium Disasters*. New York, Oxford University Press.

Brown, K. [2019](2020). *Manual for Survival: An Environmental History of the Chernobyl Disaster*. New York, NY: W.W. Norton & Company.

Dobraszczyk, P. (2017). *The Dead City: Urban Ruins and the Spectacle of Decay*. London - New York: I.B.Tauris.

Huyssen, A. (2003). *Present Pasts: Urban Palimpsests and the Politics of Memory*. Stanford, CA: Stanford University Press.

Irving, M. (2003). *A new tomb for Chernobyl*. Domus, 863. pp. 95-99.

Young, L. (2016). *An Atlas of Fiducial Landscapes: Touring the Architectures of Machine Vision*. Log, 36. pp. 125-134.

Schmid, S.D. (2015). *Producing Power: The Pre-Chernobyl History of the Soviet Nuclear Industry*. Cambridge, Massachusetts: The MIT Press.

Virilio, P. (2006). *Chernobyl twenty years later*. Domus, 891. pp. 161-162.

Whitehouse, T. (2019). *How Ruins Acquire Aesthetic Value: Modern Ruins, Ruin Porn, and the Ruin Tradition*. Cham: Springer - Palgrave Pivot.

Will, T. (2020). *Heritage as a Resource*. Domus, 1044. pp. 12-15.

View of the Musée d'Orsay from the Passerelle Léopold-Sédar-Senghor over the Seine. Author: Luigi Mandraccio (2019).

# GARE/MUSÉE D'ORSAY TO FILL, TO BALANCE

*Luigi Mandraccio*

1. Original version in French: «Nous vivons – nous survivons artificiellement – sur nos anciennes avant-gardes. Soixante-cinq ans de digestion, c'est peut-être nécessaire à une population fatiguée, c'est trop pour des créateurs. De quoi vivrons-nous demain, sur quelles avant-gardes établirons-nous l'architecture que la société réclame? Nul ne le sait, nul architecte n'en est encore le messager. Et ce n'est certes pas le musée d'Orsay qui en sera le signe. On peut affirmer au contraire que si ce musée présente un quelconque intérêt architectural, c'est parce qu'il traduit à merveille, avec le talent certain de Gae Aulenti, l'incohérence absolue de notre monde de la représentation.» Claude Parent, "Nécessité des avant-gardes", *L'architecture d'aujourd'hui* 248, 1968, p. LXXI.

«We live – we artificially survive – on our old avant-gardes. Sixty-five years of digestion may be necessary for a tired population, but it is too much for creators. What will we live on tomorrow, on what avant-gardes will we build the architecture that society demands? No one knows, no architect is yet its messenger. And it is certainly not the Musée d'Orsay that will be the sign of this. On the contrary, we can say that if this museum is of any architectural interest, it is because it conveys wonderfully, with the undoubted talent of Gae Aulenti, the absolute incoherence of our world of representation»[1]. The Claude Parent's statement – part of an elegy to radical thought – qualifies the Musée d'Orsay as a perfect device for controlling and presenting a rich and heterogeneous art collection. The challenge is managed not only by enhancing the works of art on display but also by considering the existing buildings as art works. The wise balance of the intervention creates an extraordinary and unique overall value. The establishment of the museum is a success since the transformation of the ancient *gare*. This is evidenced by decades of service as one of the main cultural tourist destinations in Paris. The project managed a complex scenario, in which there are the same "inconsistencies" s in Parent's extract. Since the ancient station is at the center of this story, it is necessary to start from how it was conceived.

"Gare d'Orsay" derives from the toponym of the area within which it was built. In fact, that stretch of the Seine's left bank was known as "Quai d'Orsay." *"Quai"* in French means "quay" (or "pier"), referring in this case to the embankment of the Seine. The name "d'Orsay" is due to Charles Boucher, lord of Orsay, mayor of Paris (1700-1708), who built the first version of this *quai*. The transformation of the ancient wood bank into an equipped structure was the beginning of the urbanization process of that part of the left bank of the Seine. By the end of the 18th century, this process development generated a complex urban area structured by streets, mansions and more straightforward buildings, related to the functions of the river port along the *quai*. At that time, the area of the future Musée d'Orsay was used as a temporary deposit. When the Ministry of Foreign Affairs – which had been designed by Napoleon Bonaparte at the beginning of the 19th century – had to be moved due to some problems in the construction of the foundations, the project was modified by Jacques Lacornée to be the seat of the Court of Auditors and the Council of State. The Palais d'Orsay, operating from 1840-1842, was a complex but significant building. In 1871, the palace was set on fire during the events of the Paris Commune and turned into a ruin, partly returning to its origin as a bush.

Despite many hypotheses, in 1897, the Compagnie du Chemin de fer de Paris à Orléans obtained permission to build a new railway station. The program was to extend the line from the Gare d'Austerlitz by underground to the Seine and then, following its course, to the point where the Palais d'Orsay was, opposite the Jardins des Tuileries. The new head station was created not only for bringing the line closer to the center of Paris, but also for the 1900 World's Fair. Both foreign delegations and visitors arrived in Paris there. It was such a prestigious task.
The Compagnie du Chemin de fer de Paris à Orléans prepared a detailed preliminary project of the railway, the station, and the annexed hotel. The aim was to set up the intervention's main characteristics, especially from the technical point of view.
The new electrified railway line ran through tunnels connected to the underground level within the station's large volume. On the surface, the buildings occupied approximately a rectangle of 170 by 75 meters – enclosed by Quai d'Orsay, Caisse des Dépôt et Consignations (1816), Rue de Lille and Rue de Bellechasse.
The preliminary project was about three adjacent main buildings following the longitudinal axis of the area: first, the large vestibule (two side forepart, called "pavilions" and a lower central part) with the primary services for passengers differentiated into "Grandes Lignes" and "Banlieue"; second, the area of the railway tracks under the large glass vault; third the multi-story building, tall and narrow, on Rue de Lille. Another building was located in front of the head of the large vaulted volume on Rue de Bellechasse.
The station worked like this: passengers entered from the Quai d'Orsay and then proceeded to the preliminary departure activities and finally descended to the platforms. Arriving by train, people proceeded along the platforms, then climbed the stairs and elevators located at the end of the large glass nave, waiting for baggage to be delivered and finally exiting onto Rue de Bellechasse.
The Compagnie launched the competition to give "architectural character" to the preliminary technical project. Victor Laloux's proposal won. He was a successful architect – winner of the Prix de Rome for Architecture – and author of the Tours' railway station (1896-98) and other public facilities in France.
His project arose thanks to some decisive choices which perfected the Compagnie's approach. In the area of the tracks, the iron and glass vault of the central nave was developed further. Nevertheless, it was hidden by raising the building's part towards the Seine with an attic structure covered in slate. The whole elevation on the Quai d'Orsay is well balanced, resulting as an elegant and monumental front that supports the building's public role within the city. On the opposite side, towards Rue de Lille, Laloux employed the tall and narrow hotel to hide the large glass roof. The same logic still guides the building on Rue de Bellechasse, which precedes and hides the glazed tympanum of the central bay. On the upper floors, it was part of the hotel, while on the ground floor, including the inner courtyard, it was the exit from the station, under the large canopy.

Musée d'Orsay interior panoramic view.
Source: Wikipedia Commons.

The large vaulted iron and glass roof – superfluous but powerfully symbolic – contrasts with the idea of hiding it on every front.
Not surprisingly, these contradictions were common at the turn of the nineteenth and early twentieth centuries. However, that solution probably increased the sense of wonder that one felt upon entering that space so massive and monumental as it was difficult to imagine from the outside.
Laloux's Gare d'Orsay was linked with the context on every scale. Thanks to this relations system, it immediately had a central role within the city and the Parisian urban fabric, especially as a new functional node. Its design solution was articulated by giving a different architectural character to the two main parts – station and hotel. The station followed the tradition of the significant monumental public buildings in Paris. The hotel (370 rooms) was instead in tune with the surrounding urban fabric, of which it assumed scale and decoration.

The station was officially opened on July 14, 1900.
Then, it was abandoned gradually between the 1950s and 1970s, after having been the head station of the line of the Compagnie du chemin de fer de Paris à Orléans for thirty-nine years. Considering the evolution of railway traffic, it became too short.
So, it was downgraded, starting in 1939, to suburban traffic only, while the main line's traffic stopped again at the Gare d'Austerlitz. At the end of the 1970s, the railway station was wholly superfluous and was transformed into a metro station.
However, the Gare d'Orsay was recognized and protected as a monument – since 1973. The hypothesis of demolition, supported by many during the 1960s, including Le Corbusier, was averted: they saved it.
Almost simultaneously with the protection plan, the idea of transforming the station into a museum was born mainly thanks to

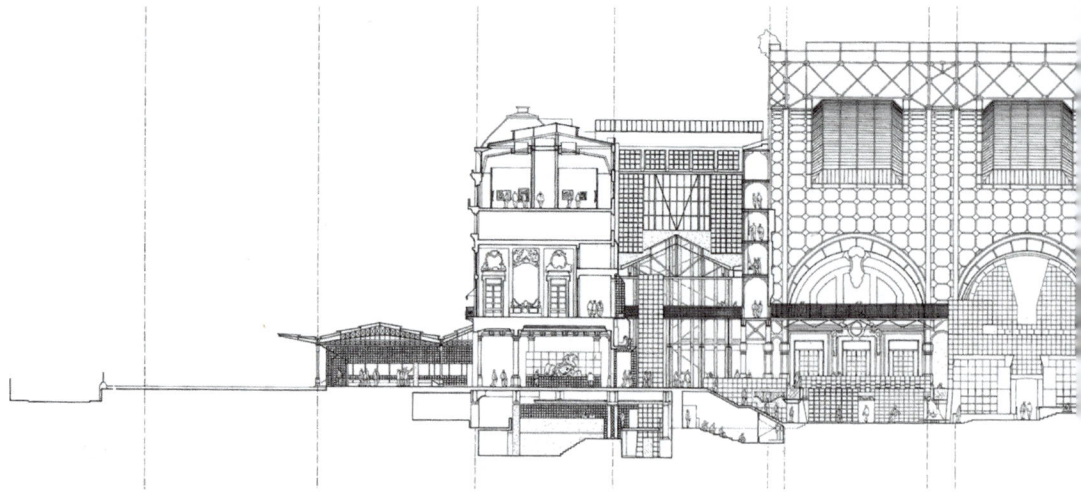

Gae Aulenti, Musée d'Orsay, longitudinal section along the avenue (1982 version). Courtesy of Archivio Gae Aulenti.

Michel Laclotte. He is an Art historian from the École du Louvre, later becoming one of the leaders of the Musée du Louvre. Since 1973 the Directorate of the Museums of France studied the possibility of creating new exhibition institutions. The goal was to decongest the other museums and enhance the artistic heritage – creating a unique, organic layout, especially for the Impressionists. Finally, it was decided to make a museum dedicated to all the arts of the nineteenth century. The museography program was drawn up by Laclotte between 1977 and 1982, taking into account all possible artistic expressions and a section for architecture and urban planning. The time frame was set between 1848-1850 and 1905-1908, thus marking the watershed with the Louvre Museum and the National Museum of Modern Art.

The objective constraints imposed by the new structure – which would have counted an exhibition area of about 47,000 square meters – were then taken into utmost consideration. There was an outstanding equilibrium, in fact, between the museum program and the existing structure to be restored: a virtuous compromise that represented an enrichment for the overall project.

To design a museum always means to reach the right balance between container and content: the quality of the container must be maximum to contribute to the enhancement of the works of art, but at the same time, it must not disturb the experience of the users. Then there is the compromise between the presence of a broad public and the need for concentration to appreciate the works of art and the compromise between security and fluid circulation in the museum spaces.

There are further issues in transforming a station into a museum. The problems arise from the almost antithetical nature of the two functions. The spaces of the station are so open and expansive. At the same time, the museum needs a multitude of intimate spaces, in which the exhibition area is maximum, and the experience is comfortable. Two competitions addressed this ambitious program. Six groups of French architects were invited to submit a proposal in

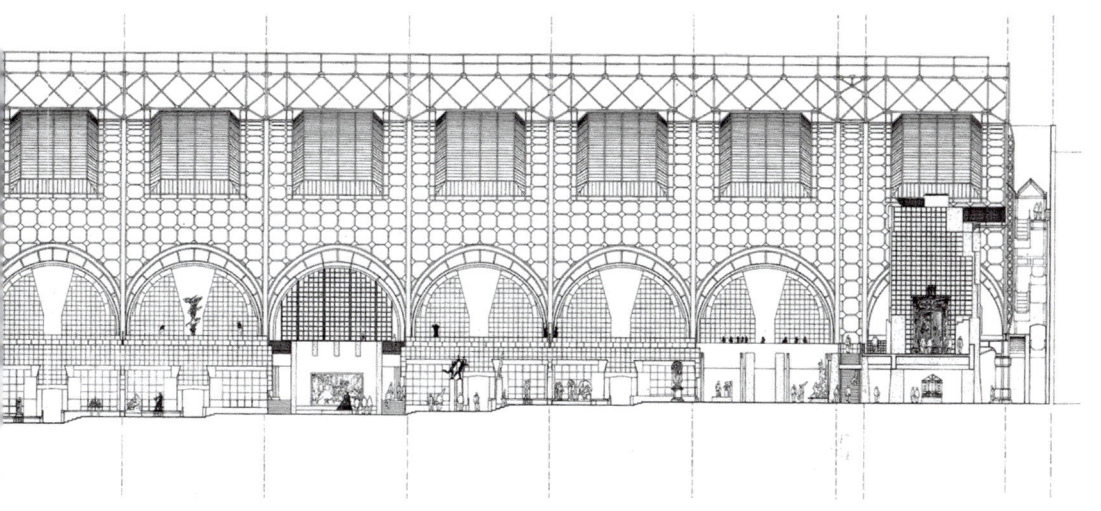

October 1978. Then the second phase was an "interior architecture" competition, launched in May 1980.
The ACT Architecture agency won the first-degree competition by interpreting contextual constraints in the best possible way. The goal of maintaining and enhancing the original building first concerned the large vaulted nave. Most of the new spatial devices for the display of works of art were planned there. The ACT agency's proposal was an overall system in which distribution and exhibition spaces are developed simultaneously and effectively intertwined. This made it possible to satisfy many of the museum's needs in a balanced way. The longitudinal axis of the buildings became the matrix of all the main paths of the project. First of all, the central "avenue" along the central nave's entire length was an ascending path starting from the original height of the tracks.
On the sides of this broad avenue – that is, in the vestibules of the old station and the hotel building, but also occupying part of the central nave – a new intermediate floor was designed to increase and better organize the various exhibition spaces. The exhibition halls are located, at various levels, on the side facing the Seine (the most impressive and spacious) and on the side facing Rue de Lille. The station's wholly original state had to be preserved in one area, but unfortunately, this is not the case today.

The intervention of Gae Aulenti – winner of the second-degree competition for the museum's internal architecture – reworked the general project of the ACT agency, according to the development of the museum program. The aim was to give it a homogeneous and coherent definition. Her approach was neutral for the ideologies of restoration and reuse, focusing on functional and formal needs. The fundamentals were kept in the final version. The rotation of the functional axis of the buildings – moving the accesses from the Quai d'Orsay to the Esplanade on Rue de Bellechasse – remained. Still, the details of the project resulting from this decision were improved. The staircase's elliptical structure between the entrance and the avenue was eliminated, creating a straighter and more

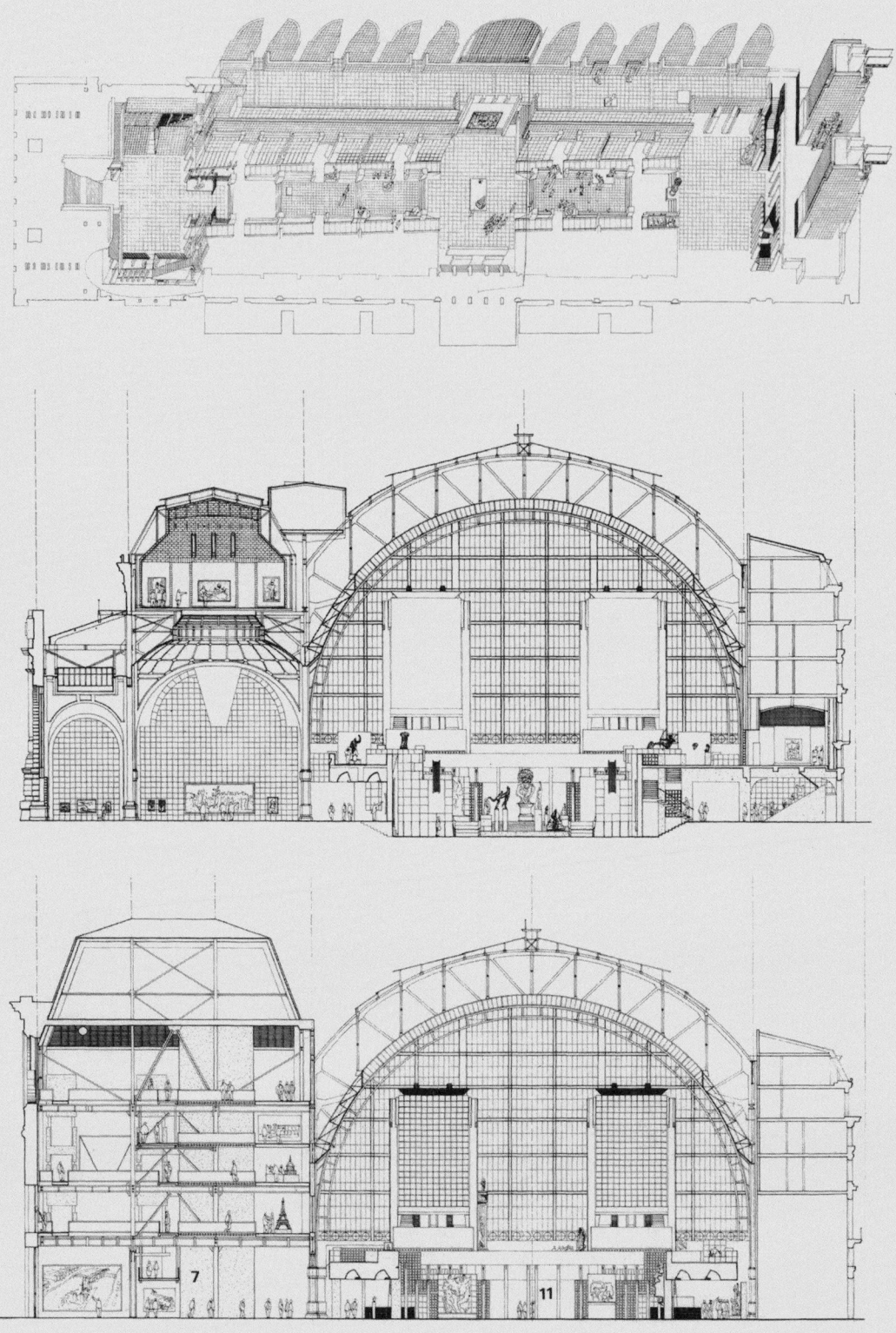

linear staircase instead. The windows provided along the entire longitudinal development on the side of the "terraces" (where the new attic advanced to the great nave) were also eliminated. Moreover, the end of the central volume was also modified, eliminating both the large monumental "double arch" planned near the large glazed tympanum, and the arena imagined at its feet.
In their place, Gae Aulenti adds two towers as special display devices, reachable via two linear stairs. The strong sense of direction established by the longitudinal axis could have made the system of the side rooms marginal. Therefore, in the final project, a network of paths overlapped the longitudinal axes with a series of transverse lines, to make everything more dynamic and balanced. Alongside these innovations, the final project by Gae Aulenti worked very intensively on all the details of the spaces and the exhibition devices. On the one hand, it established the fortune of inserting the new museum inside the old station, in terms of respect for the existing buildings and continuity with the structures' elements. On the other hand, the result was a complex museum space that is fully functional and very comfortable for visitors.
Gae Aulenti analyzed the original building as a contemporary object, without history, in all its parts: the geometry, the iron structure, the stone walls, the decorations, and its organization. The analytic process, through decomposition and fragmentation, was placed in systematic opposition and not in continuity. However, since the basis of the new language was precisely the original elements, the final result does not break but is cleverly inserted. Every form, every constructive detail, was never like a definitive and finished presence, but like a joint of different parts that were broken down and then reassembled. Among these, the study of lighting and natural light control played an important role.

In the Musée d'Orsay, each place is clearly defined and results from an intense design experience throughout the process. A long path – mainly considering Victor Laloux's station – in which we see the same attitude towards the specific context – at first the city, the urban landscape, and then the buildings in a progressive descent of scales. Here "respect" is a complex and not evident attitude.
The great project phases all establish dynamic balances that concern architecture and, ultimately, men: an extraordinary museum, the result of non-ordinary people and conditions.

Bibliography

L' Architecture d'Aujourd'hui 248, December 1986.

Petranzan, M. (2002). Gae Aulenti. Milano: Skira.

Zardini, M. (1987). Gae Aulenti e il Museo d'Orsay. Milano: Electa.

# TOWARDS A BACK-SIDE CULTURE.
# FROM ARCHITECTURAL TO PLANETARY SCALE: DESIGN ATTEMPTS TO OVERCOME THE OBSESSION OF THE FRONT.

*Francesco Garofalo*

## Back of House: People, Architecture, City

In architecture, the theme of the *back-of-house* is recurring and catalyses critics and designers' attention. It is common practice among architects to develop multiple attempts to hide the *B-side* of buildings to the point of institutionalizing a frontal approach to architecture, opposed by rarefied attempts to develop a "rear aesthetic". The front-back hierarchy well describes the cultural phenomenon of the façade, where space and visibility is granted to everything deemed worth to be displayed, while the back becomes the hidden condition, where to concentrate what needs to be concealed from the citizen-tourist-consumer 'sight.
The service exits located on the rear counterbalance the atriums on the *façade*; the kitchens are opposed to the dining rooms, the parking lots to the hotel lobbies. Starting from the building, the persistent condition of hiding the back can be read by anthropomorphic analogy in the human body to the point that, as Spyros Papapetros argues, "the veneration of the face and the repression of all that is rear are to be considered the pillars of all literature at the basis of architecture" (2014). Following this argument, the front-back dualism can also be identified in urban culture and in the morphology of cities: both as a voluntary intention and as a consequence of a hierarchical planning of the front and back.

## The Back of Urban Society: Security Planning

Corresponding to the obsession of the front is the obsession of hiding everything that is marginal, everything that does not follow the dictates of a consumer society, everything that does not respond to an ambiguous and instrumental conception of decency. Hostility towards the rear corresponds to hostility towards the marginal, "the quintessence of neoliberal securitarianism" (Bukowsky, 2019). Safety planning, aimed at eliminating any source of danger, even if only perceived, has its foundations in the "social construction of insecurity" (ibid.). The ostentation of control is not only exhibited, is flaunted, as in the case of the countless CCTV cameras that monitor stations, roads, parks, becoming themselves consequently, symbols of danger; yet: if there are, there will be a reason. The display of control reaches its maximum in "hostile architecture", a design strategy that produces spaces and furniture based on their capacity to exclude. This discipline produces, as defined by critic Frank Swain, the perfect anti-object, "a strange artefact, defined far more by what it is not than by what it is" (Swain, 2013).

Sitting on Spikes. Author: Openfabric (2020)

Hostile architecture, with its anti-seat-spikes, anti-bivouac-curves, anti-crowd-armrests, anti-skate-curbs, produces the design tools to drive away "those who do not have the privilege of accessing wealth"(Scaffidi, 2020). The built space materializes the distinction between what can be done from what cannot, between who can stay and who cannot. In this way, groups considered inappropriate are moved away, displacing them elsewhere, marginalizing communities and fragmenting cities. Unlike architecture, which must necessarily accommodate a back-of-house, indispensable for the functionality of the building, on an urban scale, the urban planning of exclusion makes use of the spatial indeterminacy of the city, tending to avoid facing the topic directly. The question of the rear, both morphological and cultural, the latter identifiable in the most fragile fringes of society, is avoided; decisions on the matter are postponed, delegating a possible solution to other portions of the city. By delegating the decision-making role to a vague "other", a process of de-responsibility is triggered accentuating processes of urban polarization and segregation of its communities.

## The Back of the City: Suburbs and Infrastructures

The zoning produced by a front-back view of the city and its inhabitants accentuates the urban phenomena of existing margins and generates new ones. If the city centre becomes the ideal place for touristic consumption, the suburbs become the urban extension of the back, places of concentration of functions, spaces and dynamics that must be kept away from the eyes. Districts defined by logistics and industrial areas, spaces for the most fragile ones, dislocated by the safety-driven urban planning. The urban-backs are leftover spaces, complex and articulated, non-planned areas of accumulation of support functions to urban life and fertile grounds for proliferation of informal activities. The industrial and logistical areas, accessible only to workers, generate an impenetrable border, which follows dynamics like those of the blind facades. The back of commercial areas is traditionally occupied by ample supporting operational spaces, areas for loading and unloading of goods and for the accumulation of waste, similarly, on an architectural scale, to the rear side of restaurants. At urban scale, by delegating the resolution of the problem to the neighbouring areas, the urban-back becomes a non-designed space, an unavoidable by-product of the urban metabolism.

## Towards a Rear Culture: 3 Test Cases

A holistic reinterpretation of the rear emerges as key to approach the back-of-house conditions as a complex, architectural, urban and social phenomenon. It offers an essential view to interpreting contemporary urban condition, serving as a critical observation tool able to develop responsible planning and ultimately creating the conditions for a "rear culture". Similarly, to what happens with waste, the residual spaces (leftovers) generated by the back can be seen as places to be re-interpreted, places on which to project new urban roles and new identities. It is essential to invert the cultural paradigm of the back as a taboo, as a space to hide, offering a reinterpretation of these marginal areas as spaces to be revealed because of their intrinsic characteristics and underestimated potentialities. The opportunity arises to capitalize on these specific characteristics, such as their ability to attract all what is marginal, their propensity to create the conditions for non-traditional activities and to activate both social and natural informal dynamics. The traditional poor maintenance of these spaces, which sometimes extends to total decay, offers a fertile ground to be colonized by pioneer biodiversity, new urban habitats with great connective potential. The lack of control makes these places, spaces for the proliferation of non-traditional dynamics, phenomena which are repressed by the control-obsession carried out by security-lead planning, in more privileged portions of the city. This paper aims to present some attempts – more or less speculative – to reinterpret these areas at different scales, through three design exercises developed by Openfabric, a landscape and urban design practice founded in 2011 in Rotterdam, the Netherlands.

Loading and unloading exposed.
Author: Openfabric (2021)

A new proximity between citizens and their port. Author: Openfabric (2017)

**Back-side Carlo Felice / Genoa**
*Spotlights on the daily life*

We are witnessing the economical exploitation of many urban public spaces, where "it happens more and more often to see wonderful historical squares invaded, indeed hidden, by stages, scaffolding, spotlights, chairs for spectators, barriers, sports equipment" (Settis, 2017) and such spaces are «closed to the non-paying public» (ibid.). The need emerges to restore relevance to everyday life. If the traditional square is often perceived as the space for the (large) event, it is essential to rebalance the equation in favour of the "non-event," giving central stage to all what is "usually unnoticed" (Perec, 1975). In a redevelopment project of the space adjacent to the Carlo Felice theatre in Genoa, Openfabric proposes the spectacularization of the rear. Theatres, like many other types of urban architectures, are characterized by a clear difference between the front – Piazza de Ferrari entrance, the main square of the city – and a rear, a logistic space for loading and unloading goods to support theatrical activity – Largo delle Fucine. The project aims to give new importance to the background life of the shows: the intense logistical activity at the back where large trucks transport portions of scenography, transferred inside through a large service entrance. Through a paving and lighting project, Openfabric proposes to transform this obscure infrastructural space into a lively portion of the city to look at, with interest. A lighting system, inspired by stage-lighting techniques, reinforces the intention to make the loading and unloading activity a "show" to be watched every day, an urban stage where trucks, sceneries and operators become the protagonists of the urban scene. The back remains an operational backside, but is treated as a place to be revealed, and the activities, usually considered to be hidden, become urban performances to be displayed.

**Archipelago of Knowledge/ Rotterdam**
*From barrier to space*

Industrial and logistic areas generate urban barriers which, in most cases, are not accessible to the population. The activities themselves that find space in these areas are usually irreconcilable with the presence of humans, and often have negative effects on neighbouring areas. The question of the rear therefore emerges, both as a physical margin and as a condition of widespread degradation, with various declinations, depending on the industrial and logistic activities in question.

Archipelago of Knowledge is a speculative vision on the future of the northern part of the port of Rotterdam. Openfabric proposes here a clear differentiation between port areas whose activities are incompatible with public life and areas that are potentially accessible, where new conditions can be created for a virtuous proximity between citizens and their port.

The strategy defines the inaccessible port clusters and proposes their spatial isolation. Through the excavation of a new canal, these areas become islands, for the exclusive use of port workers, and at the same time transforms the border, that was once impenetrable, in a new accessible space. The canal re-establishes direct contact between the city and the water which was largely lost in the development and growth of the port. The canal works as a water

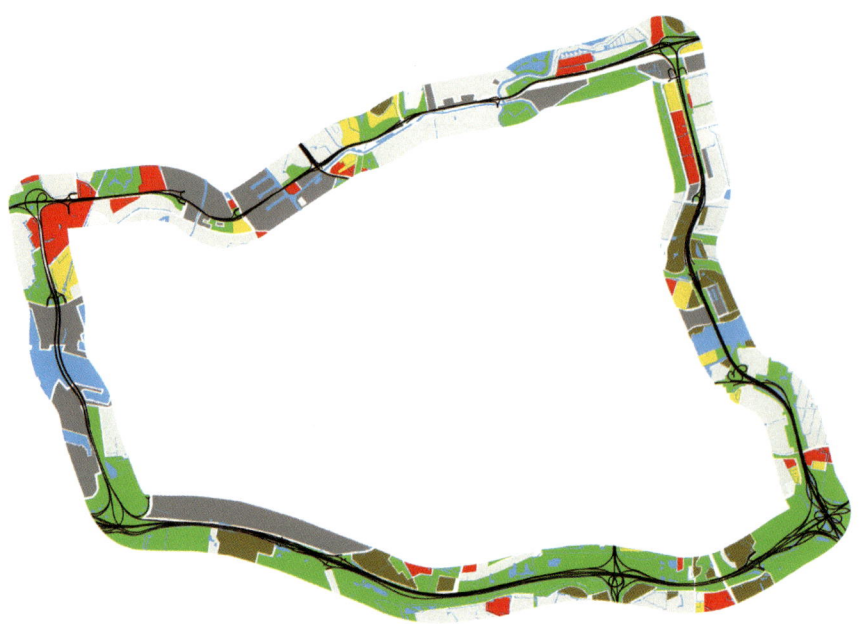

Spatial and programmatic complexity of infrastructural leftovers. Author: Openfabric (2016)

transport infrastructure while it generates a waterfront, a new catalyst for public life. Here the border becomes a space for sharing, re-establishing a relationship between citizens and industrial activities – made visible and spectacularised – and between citizens and water – made accessible and programmed.

Road infrastructures, not unlike railway ones, when they cross urbanized territories generate a series of leftover spaces. Due to their shape and size, and the proximity to the infrastructure as an environmental source of disturbance, these places often don't have a specific function. The sole purpose of such leftovers is to create a protective buffer between infrastructures and residential areas. These fragmented and highly inaccessible areas are, in common perception, alien to the city: the perfect context for J.G. Ballard to set his novel Concrete Island (1985), where the protagonist Robert Maitland is trapped following a road accident in one of these interstitial spaces between infrastructures, without being able to get out. The linear and continuous nature of the infrastructure gives these spaces the potential to be transformed into highly connective landscapes. In defining a strategy for the Rotterdam ring-road, Openfabric proposes to capitalize the ecological potential of these low-maintenance spaces, rendering them accessible and proposing new programmatic layouts.

The strategy, based on the need expressed by the client to rethink the infrastructural landscape over a few decades-long time scale, cannot ignore the rapid evolution of the means of transport. Mobility trends show a drastic reduction of environmental sources of disruptions, such as noise, $CO_2$ and particulate dust. In proposing a system of large public spaces, Da-Ring strategy articulates the

**Da-ring / Rotterdam**
*From back to front*

possibility – considering the expected technological developments – of inverting the paradigm where mobility infrastructures are generators of urban backsides, noisy, polluting and inaccessible spaces from which to stay away.

The result is a system of parks unfolding along the ring road: this resulting landscape is organized into a new hybrid infrastructure, where the ecological role is strengthened, and underused spaces are revalued for the benefit of the community, creating the conditions, over time, to transform such urban back-side into a new front.

## From Leftovers to Opportunities: Towards a Back-Side Culture

These three design elaborations highlight an updated point of view on the phenomenon of the rear and propose possible new scenarios. The multiscalarity of these examples reflect the magnitude of the rear-dynamics and can be deployed to observe, interpret and design forgotten part of the urban fabric. The base-line understanding is to acknowledge the city-scale back-of-house and to investigate opportunities beyond the architectural scale, opening-up possibilities for a paradigmatic change of their role in urban dynamics: from leftovers to opportunities.

Through these design experiments, the urban-rear be considered as a valuable landscape, and at the same time, the system of these spaces, networked, becomes infrastructure. If we see the "landscape as infrastructure" (Belanger, 2016) these spatial by-products of urbanization are "connective tissues and the circulatory systems of modernity. In short, these systems have become infrastructures" (Edwards, 2002).

By extending this viewpoint to the global scale, for instance from a Planetary Urbanization (Brenner and Schmid, 2012) perspective, the proposed framework lends itself to understand consumption and extractive dynamics as opposed. The front-back dualism is recognisable on a planetary scale, where privileged urban areas exist due to exploitation of resources in the "global rear", supporting a lifestyle that is as elitist as it is unsustainable.

## Bibliography

Ballard J.G. (1985). Concrete Island. New York: Knopf Doubleday Publishing Group.

Belanger P. (2016). Landscape as Infrastructure. London: Routledge.

Brenner N, Schmid C. (2012) "Planetary urbanization"; in Gandy M. Urban Constellations, Berlin: Jovis, pp. 10-14.

Bukowsky W (2019). La buona educazione degli oppressi. Piccola storia del decoro. Roma: Edizioni Alegre.

Edwards P. (2002) "Infrastructure and Modernity: Scales of Force, Time, and Social Organization in the History of Sociotechnical Systems", in Modernity and Technology, Cambridge: MIT Press.

Papapetros S. (2014). "Building from behind: Architecture and the aesthetics of dorsality", in Area, n. 112 Beauty of Built.

Perec G. (1975). An Attempt at Exhausting a Place in Paris. New York: Wakefield Press.

Scaffidi L. (2020). "Architettura ostile. Milano è smart o unpleasant?", in Menelique Editoriale n. 2. La città muta.

Settis S. (2017, July 12) "La piazza che diventa location è morta", Emergenza Cultura. Retrieved from www.emergenzacultura.org/2017/07/12/salvatore-settis-la-piazza-che-diventa-location-e-morta/

Swain F. (2013, December 5). "Designing the Perfect Anti-Object" in Medium. Retrieved from www.medium.com/futures-exchange/designing-the-perfect-anti-object-49a184a6667a#.kzs5zujg3

Web References:
Openfabric, Rotterdam, NL
www.openfabric.eu/

# #2 RESEARCH

«The transposition of landscape and infrastructure outlines how the field of landscape – once architecture's surrogate – can break free and exploit its **affiliation to ecology, engineering, and geography by reengaging large-scale planning and reimagining small-scale surfaces and materials** through a reform of existing urban infrastructures and a projection of new ones».

- Pierre Bélanger (2017), *Landscape as Infrastructure*.

# READINGS AND CASE STUDIES

**A preliminary sample: 4 binomials, 28 factsheets**

The relationship between the multiple forms of contemporary landscapes and the various declinations of infrastructure allows the construction of a heterogeneous framework of projects, which are collected and analyzed here as research case studies. These are projects in progress or recently completed, selected on a global scale. Although the focus is on Europe, the case studies of this research extend from the Mediterranean basin to the Atlantic coasts of the United States and on to South America, the Korean shores, China and New Zealand.

The 28 selected projects are organized into four thematic groups, which refer to four strategic binomials. Each binomial addresses the theoretical-strategic relationship between the infrastructure and another element considered crucial and analyses, through the selected projects, the possible impacts in terms of design, spatial performance, functional programme, formal aspects of the architecture, etc. The 4 binomials taken into consideration are identified conceptually as follows: **infrastructure *and* nature (INA)**, **infrastructure *and* public space (ISP)**, **infrastructure *and* industry (IIND)**, **infrastructure *and* infrastructure (IINF)**.

Each project is accompanied by a summary factsheet containing the main data of the project, its geographical location, the state of the art, the chronological phases of its realization and the type of project (public, private, competition of ideas). By means of a number of key words specially selected and attributed to each project, the factsheets are proposed as a study tool: they offer a constantly implementable overview of projects to refer to and from which to derive contemporary approaches and practices to be understood and replicated.

Ranging, for instance, from the complex masterplan of La Confluence, Ilot A3 in Lyon, France, to the Seoullo 7017 Sky Garden in South Korea, from the Metro Cable in the popular neighborhood of Caracas called San Agustín, to the system of redevelopment projects for disused railway yards in Milan, the 28 case studies shift between the different scales of design and dimensions of intervention. By doing so, these case studies witness the versatility of the projects addressing the relationship between landscape and infrastructure and, at the same time, show the diversity of the stakeholders (local authorities, investors, public associations, citizens, private companies, etc.) involved in the design and realization processes.

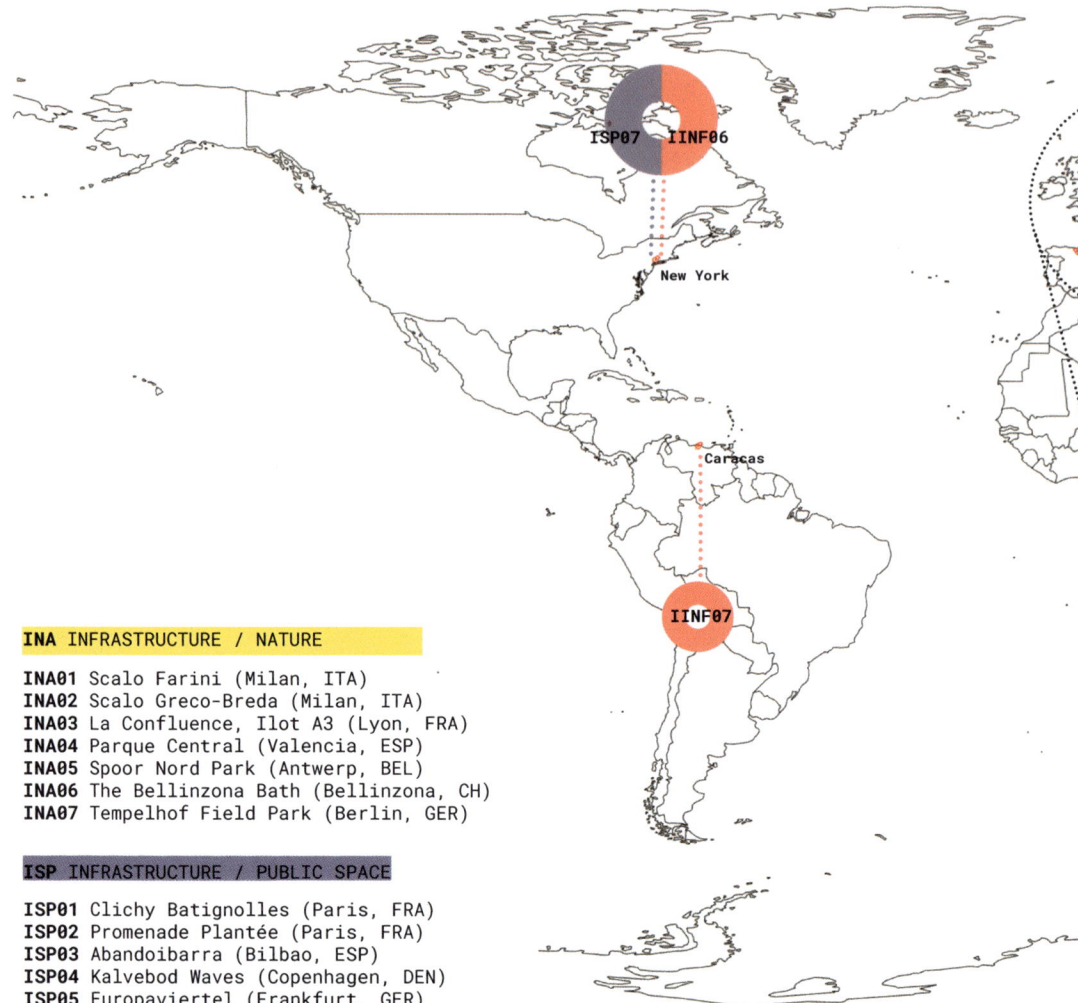

### INA INFRASTRUCTURE / NATURE

**INA01** Scalo Farini (Milan, ITA)
**INA02** Scalo Greco-Breda (Milan, ITA)
**INA03** La Confluence, Ilot A3 (Lyon, FRA)
**INA04** Parque Central (Valencia, ESP)
**INA05** Spoor Nord Park (Antwerp, BEL)
**INA06** The Bellinzona Bath (Bellinzona, CH)
**INA07** Tempelhof Field Park (Berlin, GER)

### ISP INFRASTRUCTURE / PUBLIC SPACE

**ISP01** Clichy Batignolles (Paris, FRA)
**ISP02** Promenade Plantée (Paris, FRA)
**ISP03** Abandoibarra (Bilbao, ESP)
**ISP04** Kalvebod Waves (Copenhagen, DEN)
**ISP05** Europaviertel (Frankfurt, GER)
**ISP06** Seoullo 7017 Skygarden (Seoul, KOR)
**ISP07** The High Line (New York City, USA)

### IIND INFRASTRUCTURE / INDUSTRY

**IIND01** High Speed Train Station (Logroño, ESP)
**IIND02** Stratford City (London, GBR)
**IIND03** King's Cross (London, GBR)
**IIND04** Zuidas (Amsterdam, NED)
**IIND05** Rotterdam Central Station (Rotterdam, NED)
**IIND06** Yangpu Riverside (Shanghai, CHI)
**IIND07** Auckland Waterfront (Auckland, NZL)

### IINF INFRASTRUCTURE / INFRASTRUCTURE

**IINF01** Scalo Porta Romana (Milan, ITA)
**IINF02** Scalo Lambrate (Milan, ITA)
**IINF03** La Antigua Estación (Burgos, ESP)
**IINF04** Central Pasila (Helsinki, FIN)
**IINF05** Europaviertel (Stuttgart, GER)
**IINF06** Hudson Yards (New York, USA)
**IINF07** Metro Cable (Caracas, VEN)

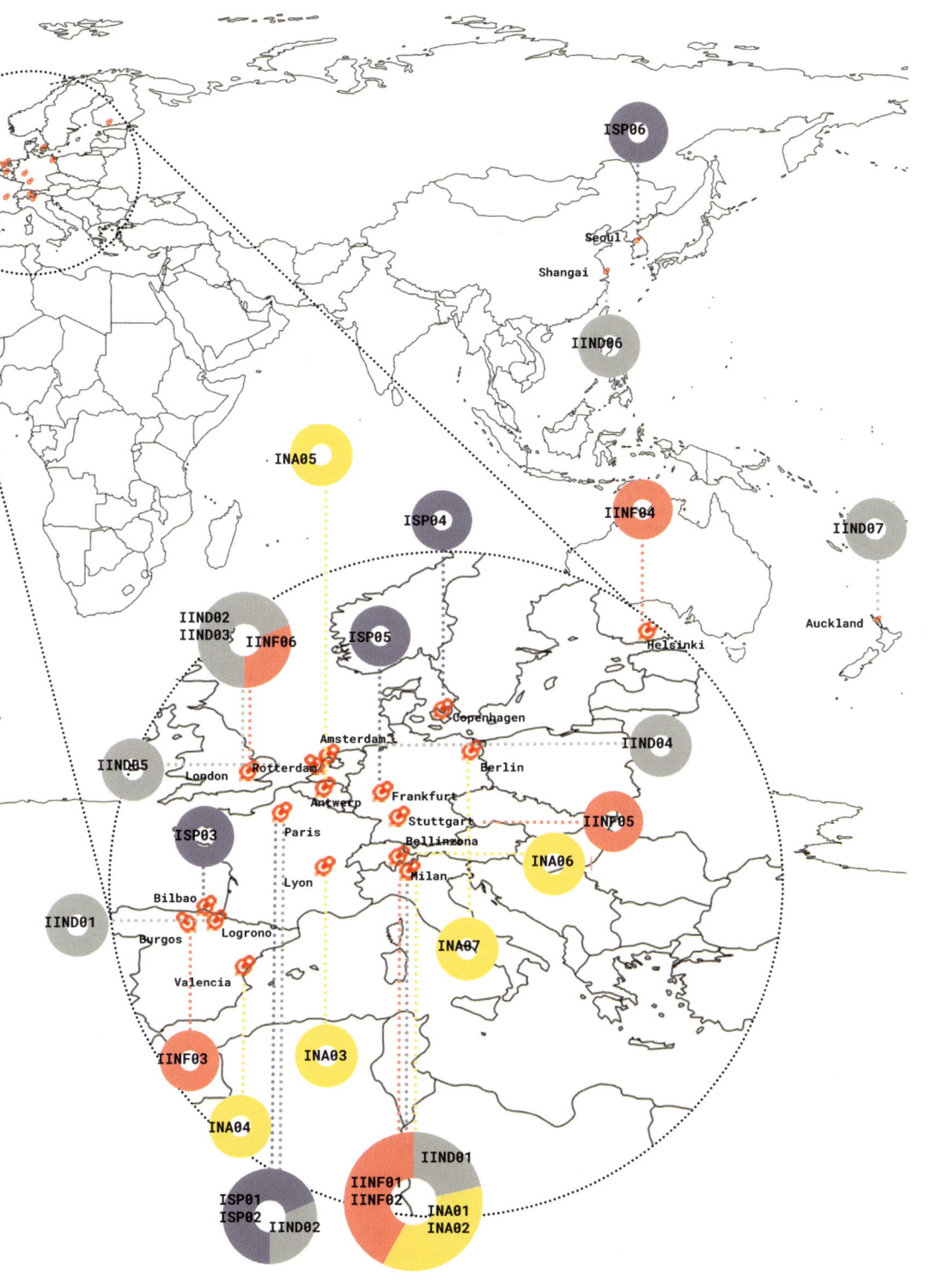

# #INA01 | SCALO FARINI

Milan, ITA
status | 2019 - ongoing
typology | COMPETITION

The Farini railway yard straddles very different parts of the city of Milan, including the more central ones, around Porta Nuova, to those of the peripheral Bovisa. The internal part of the airport contains artifacts of different categories: from sheds for the shelter of trains to buildings connected to customs. The railway yard is crossed by underground paths, while on the surface, some pre-existing railway lines remain. The railway yard, consisting of an area of 404,000 sm, is surrounded by two important traffic systems that constitute a sort of barrier. The railway yard was the subject of the "Farini Competition" which is part of the Program Agreement signed in 2017 by the Municipality of Milan, the Lombardy Region, Ferrovie dello Stato Italiane and Savills Investment Management (today COIMA) for the redevelopment of the seven disused railway yards. Overall, the seven railway stations occupy an area of 1 million and 250,000 square meters, of which approximately 200,000 will remain as a railway station. This is the largest urban regeneration plan in Milan and one of the largest territorial enhancement projects within Italy and Europe. Through the C40 Reiventing Cities platform, FS Sistemi Urbani and COIMA sgr announced the "Farini Competition" aimed at drawing up the Masterplan for the regeneration of the disused railway yards of Farini and San Cristoforo. The Masterplans, foreseen for the railway stations of Farini and San Cristoforo, Romana and Genoa, are functional to the subsequent presentation of the Implementation Plans. The competition for the redevelopment of the Farini railway yard was won by the "Climate Agents" project, presented by the group led by OMA and including Laboratorio Permanente, Vogt Landscape Architects, Philippe Rahm Architectes, Net Engineering, Ezio Micelli, Arcadis Italia, Temporiuso.net and Luca Cozzani. The aim of the winning project is to transform the railway yard into a green area that acts as an ecological filter, in fact an equipped linear park and public spaces are designed to help purify the area and mitigate the heat island effect. The Farini railway yard will connect areas that have seen major transformations (Porta Nuova) with areas that will soon see them, such as the Bovisa axis. The image proposed by the intervention is the creation of a large park that unites two parts of the city that are now separated. The park, which should occupy at least 60% of the area, unites distant urban districts also through interventions to mend the vehicular traffic, at the service of public transport, cycle and pedestrian paths. From an environmental point of view, the proposal interprets the open space as a large park adaptable to different circumstances and uses: a system of green spaces connected but with different natures and functions, inside and outside the airport, such as urban gardens, a range of connection enhancement of the park of Villa Simonetta, the Testori park and open spaces of the Rinnovata-Pizzigoni school.

Render of the former Farini railway yard as required by the winning master plan.
Authors: OMA and Laboratorio Permanente.
Source: oma.eu/projects/scalo-farini

# #INA02 | SCALO GRECO-BREDA

Milan, ITA
status | 2018 - ongoing
typology | COMPETITION

The Greco-Breda railway station is located in a strategic urban position, especially following the interventions that involved it in the Bicocca system, aimed at connecting the city center to the suburbs. Used in past epochs exclusively to connect the historic village of Greco to the countryside, in more recent times the Greco-Breda transport line was mainly used as a freight yard, disused and then replaced by the Greco-Pirelli port which is still isolated from to the surrounding city areas. The Greco-Breda railway yard was the subject of the Program Agreement signed in 2017 by the Municipality of Milan, the Lombardy Region, Ferrovie dello Stato Italiane and Savills Investment Management (today COIMA) for the redevelopment of the seven disused railway yards. Taken together, the seven airports occupy an area of 1 million and 250,000 sm, of which approximately 200,000 will remain in railway function. This is the largest urban regeneration plan in Milan and one of the largest territorial enhancement projects in Italy and Europe. Through the C40 Reiventing Cities platform, FS Sistemi Urbani and COIMA sgr have announced the "Farini Competition" aimed at drawing up the Masterplan for the regeneration of the disused railway yards of Farini and San Cristoforo. The Masterplans, foreseen for the railway stations of Farini and San Cristoforo, Romana and Genoa, are functional to the subsequent presentation of the Implementation Plans. The winning project is "L'Innesto" signed by the team of architects Barreca & La Varra and Arup Italia, which aims to increase the residential and commercial areas of the district and to create pedestrian paths connecting the surrounding green diaphragms, such as the North Park, the Martesana Park and Lambro Park. As required by the objectives of the call, the project proposes a series of interventions on an area of 62,189 square meters focusing on mobility, in particular on the enhancement of the cycle / pedestrian area, on the improvement of vehicle accessibility and public transport. "L'Innesto" pays particular attention to the natural character of the project: the arrangement of paths and green areas and the connection to existing parks constitute the basis of the intervention and aim at improving the quality of the spaces. The infrastructural component of the Greco-Breda railway yard is integrated into the project of the new district and collaborates, together with the Greco-Pirelli railway yard, in the development of sustainable mobility and the mending of the urban fabric. The new buildings, intended for social and student residences, and public spaces with services are interspersed with green components: private and educational gardens, the avenue of mulberry trees, orchards, community gardens and woodlands. Biodiversity guarantees spatial variety and environmental richness which are also reflected in the structures built with sustainable materials through industrialized construction and technological systems with low CO2 and waste production.

Render of INNESTO, the first Zero Carbon "Social Housing" project, winner of the international competition "C40 Reinventing Cities". Authors: Barreca & La Varra Architects.
Source: barrecaelavarra.it/progetti/innesto-scalo-greco-breda/

INFRA/NATURE

RESEARCH

#regeneration #zerocarbon #socialhousing #biodiversity

# #INA03|LA CONFLUENCE ILOT A3

Lyon, FRA
status | 1998-2009 PHASE 1; 2009-ongoing PHASE 2
typology | MASTERPLAN, PUBLIC WORK

La Confluence is a former industrial district located in the center of Lyon, at the confluence of the Rhone and Saone rivers, and connected to the rest of the city by a series of transverse bridges and bouleverds. Since 1998 it has been the subject of a complex urban redevelopment plan that is transforming it into an intelligent and sustainable district aimed at respecting the original industrial heritage of the places. Located on a peninsula of 150 ha, the district has had an evolution since the mid-nineteenth century when the embankments were raised against floods, later, in 1857, the Perrache railway station was built which divided the territory in two: to the north the urbanized and modern area and to the south the industrial and port area, where two prisons were also built. In the nineties of the twentieth century, with the crisis of the old industry, the district faced a phase of decline and lost its productive connotation. In this context of progressive regeneration, the IlotA3 site is a pilot project inaugurated by the public administration in 2014 as the first batch of the second phase of investments. The masterplan project is signed by the architects Herzog & De Meuron in collaboration with the landscape architect Michel Desvigne, while a team of six emerging local and international architects were invited to design the eight buildings of the new site. Other important world designers, such as Kengo Kuma and Massimiliano Fuksas, have over time taken part in the overall intervention. Among the objectives of the masterplan are highlighted, in particular, the management of the water and energy assets through the recovery of rainwater, the preservation of biodiversity that has generated the distribution of green areas, the construction of high-performance buildings, in addition to choice of alternative modes of transport. The masterplan proposes the construction of a series of housing, offices and public structures related to trade for a population of about 16,000 units expected by 2020. All new buildings are designed to comply with pre-established environmental standards, such as consumption of less than 60 kWh/m2 per year for heating, 40 kWh/m2 for domestic hot water and 80% of energy from renewable sources. One of the most important parks is a large green lung of 14 ha, called the *Saône Park*. To the south, there is another park, *Le Champ*, of about 40 ha and made up of spaces divided by functions that penetrate the urban fabric and guarantee maximum use by the inhabitants of the new district. The relationship with water is important, which is used both as a design element for public open spaces along two rivers, and as a functional component for the water supply of the buildings under construction. From a road point of view, the project plans to reclassify the A7 highway and convert it into a city boulevard that connects La Confluence directly to the historic city center. The *pont des Girondins* will be the main artery connecting Gerland and the neighboring districts on the left bank of the Rhone.

# #INA04 | PARQUE CENTRAL

**Valencia, Spain**
**status | 2011 - ongoing**
**typology | COMPETITION**

Located in a central area of the city of about 12 hectars once occupied by railway lines and industries, the Parque Central is a project by the landscape architects Gustafson Porter + Bowman, winners of an international competition launched in 2011. The intervention involves the landfill of the still functioning tracks of the high-speed railway line that crosses a central area of Valencia and the recovery of the surface above, of about 66 ha, which will be redeveloped, giving life to the most important urban green reconversion project to date in the Spanish city. The project celebrates the landscape and culture of the Valencia region based on the position of the city settled in very different landscapes: the Turia river reserve, the agricultural plain, the natural park of Albufera and the Mediterranean basin from which it derives a central role in European trade, culture and history. These themes are found within the park through the design of different environments, some of which are obtained through elevation modeling of the ground that reflect the concave shape of traditional handcrafted ceramics. The disposal of a part of the old railway yard and the transformation of the sector into a large green lung of the city is among the citizens' objectives at the center of the discussion since the 1980s. The construction process involves an organization in three phases in order to ensure the correct autonomy and functionality of rail and urban transport. The first phase of works in the park, lasting eight years, has implemented operations capable of connecting various districts previously separated by the railway tracks, extending urban renewal in the adjacent Russafa / Ruzafa district. More than half of the area will be planted and the 23 ha park will contribute to solving environmental problems by reintroducing biodiversity, creating new public spaces, providing cultural facilities and offices, reversing physical and social segregation, improving the quality and supply of water and increasing mobility and accessibility between the surrounding residential neighborhoods. The Parque Central will consist of several areas. The Central Square located in the center, the convergence point of all the routes and the connection between the north and south sides. It will host large green areas enriched with palms and jacaranda trees and will be joined by the lake, the Mediterranean gardens and, finally, the Piazza delle Arti. Both as a structural and figurative element, the water project interprets the different uses in which it is employed in the natural landscape of the region: the sea, the Albufera freshwater lagoon and the Turia river. At each entrance to the park, visitors are greeted by a single element of water that filters into a channel that leads them inwards, putting in place a sort of continuous narration. The Parque Central also proposes an innovative strategy in terms of sustainable drainage: the series of tanks formed by the ground collect rainwater that is conveyed and recycled for city use.

Phase 1 of Parque Centraö, Valencia.
Authors: Gustafson Porter + Bowman.
Source: gp-b.com/valencia-parque-central

**INFRA/NATURE**

RESEARCH

#requalification #urban green #eco-sustainability

# #INA05 | SPOOR NOORD PARK

Antwerp, BEL
status | 2003 - 2009
typology | COMPETITION (2002)

The area (24 ha, 1.6km long) was previously a railway yard and since 1873 had been owned by the Belgian national railway company (NMBS). In the mid-1990s the old railroad depot fell into disuse and in 2001 NMBS ceased its activities there and, after a while, the City of Antwerp became the new owner of the site. Spoor Noord covers the districts of Dam, Stuivenberg and Seefhoek (325 ha) although the former maintenance yard at the centre of the regeneration (future park) covers only 29 ha. Both Stuivenberg and Seefhoek are residential areas with a complex community of immigrants, relatively few Belgian nationals, and high levels of population turnover. In terms of land use, Spoor Noord itself was mono-functional and derelict former industrial land. It created an enormous barrier between the neighboring areas. The introduction of the HST fast train (northwards from the Damplein) further damaged the continuity of the urban structure and created a new barrier in an already divided area. The release of an open public space with such dimensions in a densely-built city district offered a unique opportunity to restore the connection among the surrounding areas with a solid green lung. The aim of the municipality was to attract investment in residential and commercial land, and generate public support for the on-going regeneration process. In this situation the great advantage of Spoor Noord was the amount of open land available for public use.
Among the main goals of the project developed by Studio Associato Bernardo Secchi Paola Viganò, are enhancing sustainable urbanization and restoring ecosystems and their functions with the idea of the park as a social space: a place for many everyday activities, a city part that can specially contribute to give a clear structure to the whole city and metropolis. The proposed solution reinstates connections between three areas – Dam, Stuivenberg and Seefhoek – that were previously cut off from each other by the railway site. The main feature of the Park Spoor Noord is its dimension: a wide and simply designed lawn crisscrossed with trails that connect the various neighborhoods with each other and with the park. Gardens, sports fields, clear forests accommodate various formal and informal social practices and define multiple atmospheres. The old structures of the railway building were retained. Alongside the urban landscape park, the plan for the new Spoor Noord area was developed. This plan concentrated on the areas not included in the park, developing the idea of multifunctionality. This included the development of training clusters, a top event venue, a steam bath complex, all-in sport halls in converted hangars, new ideas for the Lobroek Dock, upgrading the slaughterhouse site and the north-south railway connection area. Together with the landscape park, this plan provided a completely integrated programme for the regeneration of Spoor Noord.
Although the programme was quite complex, the municipality adopted an implementation strategy based on small steps forward.

# #INA06|THE BELLINZONA BATH

Bellinzona, CH
status | 1967 - 1970
typology | PUBLIC PROJECT

The result of a public competition held in the summer of 1967, Bagno designed by Aurelio Galfetti, Flora Ruchat-Roncati and Ivo Trümpy, offers the first concrete manifestation of the "territorial architecture" that has characterised Ticino's architecture starting from the end of the 1960s to the most recent developments. It is a reinforced concrete structure that creates a path, or simply a six-metre high path, from the city to the river. In its rectilinear path, this elevated infrastructure connects at the same time the public space of the spa with the plain, the Castelgrande hill, the city, the mountains and the sky. Its authors describe it as "a structure that builds the landscape and organises the territory", a clear and ordering sign in the empty surrounding countryside. The functional, organisational and management aspects are addressed by subordinating them to an overall spatial vision, a single design gesture that resolves the spatial questions and operational demands of the site. The city is connected to the river through a large void by a footbridge, a structure that brings openness to the extension of the city, designed within the landscape, ready to welcome new activities. The reinforced concrete of the footbridge expresses this urban dimension, with a metal structure underneath that articulates the different functions. At zero height the system of sports facilities is distributed, including two swimming pools, an athletics track and several public open areas. A system of open-air steel footbridges detach themselves from the central axis, which functions as the primary distribution artery, and in directions perpendicular to it organise the complementary activities. Linking the city with the river, the structure takes the form of an urban expansion and gives to the context the character of an open city, projected onto the landscape, ready to receive new activities.

Aurelio Galfetti's project in Bellinzona goes beyond the creation of a single piece of architecture or urban layout, but is a commitment to the creation of a new idea of the city. Galfetti designed and partly built a number of projects between the Seventies and the Eighties: in addition to the extension of Bagno, he designed a new neighbourhood, a residential building, the restoration of Castel Grande, the new theatre and the new post office building. In this framework, Bagno is coherently linked to other buildings that enrich it on a functional level, but above all establish the limit of future urban expansion towards the river. And so, almost twenty years after its construction, the old footbridge, once conceived as a viaduct and a permeable wall, finds new connections and a raison d'être in the surrounding developments, dedicated to sports (new tennis courts, swimming pool and skating rink) and new housing models. It no longer relates to the void but to the built environment, transforming movement and circulation into a matter of planning and architecture at the same time.

INFRA/NATURE

RESEARCH

#public bath #urban development #sport center #graft

Bellinzona Bath House, 2018.
Author: Ph. Trevor Patt, Enrico Cano, Archivio del Moderno.
Source: atlasofplaces.com/architecture/il-bagno-di-bellinzona/

# #INA07 | TEMPELHOF FELD PARK

Berlin, GER
status | 2010
typology | PUBLIC PROJECT

Originally the Tempelhofer Feld was a parade ground: on weekends and on public holidays, as soon as the military cleared the site, the locals would swarm in their thousands to Tempelhof to enjoy their leisure time. At the beginning of the 1920s, Tempelhof airport was built on the site in 1936-1941 and used by the Nazis during World War II and later the city airport of Berlin. Located just south of the central district in Berlin, the site was closed for use in 2008 and as a result left a 380 acre airfield for public use. The reactivation plan of the vast park (same size as New York's Central Park) has been developing since before the closure of the airport and is concentrated on giving locals an active role in the development. A number of testing situations for long term ideas have been implemented in order to organically determine the use of the park. Many of these pioneering projects have been concentrated on sport/cultural activities which enhance the public use of the park and surrounding area to create vital links with the neighbouring community. The park is organized around six themes that guide the choices in terms of development and activities: knowledge and learning, sustainable technologies for the future, sport and health. One of the fundamental qualities of transformation of this great void is the almost total absence of programmatic compromises and, for the same reason, an opening towards the ephemeral and towards freer temporary occupations. Thus, the definition of a support of maximum flexibility and reception capacity becomes imperative, from a dimensional point of view, but at the same time in the quality of adaptation to the existing. It is therefore a holistic approach to public space planning that includes farsightedness for the development of marginal areas (between city, infrastructure and nature) and adaptive reuse of existing buildings and airport runways. The park was developed keeping the shape of the existing airport, the airport terminal is now widely used for special events: from car shows to press events and rock concerts, as well as being home to a division of the Berlin police. The rest of the building stock has largely been preserved, for example the 72 m high radar tower is still used by the German army to monitor air traffic. The park is occupied by community gardens taken care by the inhabitants of the nearby neighborhoods who use the spaces for leisure, free time, sport and community activities. Local regulations approved in the decades following the functional conversion of the park allow the conservation of Tempelhof and limit the new construction within the perimeter of the old airport. The project was conducted and financed by the city of Berlin with the design support of the German studio Raumlabor. Raumlabor's ongoing engagement on the former airfield of Tempelhof ranges from long-term planning to temporary interventions on site. it displays our role as a link between the city and its inhabitants.

INFRA/NATURE

#airport #park #public space #recycle

# #ISP01|CLICHY BATIGNOLLES

Paris, FRA
status | 2011 - 2020
typology | COMPETITION

Clichy Batignolles eco-district is today the largest building site in Paris that takes up two large city blocks contiguous with the strategic position between Haussmannian Paris and 21st-century Paris of the Clichy-Batignolles eco-district, for which it serves as bridgehead. The competition launched in 2001 concerns a large complex of buildings that host multiple functions, not only a residential one, as well as guaranteeing the presence of an open public space within an area of 54 ha, recovered from the decommissioning of the railway warehouses, it occupies 10 without counting the 6,500 sq m of private green and 16,000 sq m of green roofs. The area designed as a park was the first to be completed and delivered to the service of citizens as a natural infrastructure, putting biodiversity and sustainability first. This project, overlooking what remains of the footprint of the historic rail yards leading to the Gare Saint-Lazare, is highly complex in several ways. In particular, it is composed of a wide variety of programs, which are for the most part standing atop an artificial ground at 10 meters (32.8 feet) above the rue Cardinet. Thus, the project is designed to reconcile all the components of a mixed-use program, notably by means of superimposition, while also dealing with the connection on the street level between the ZAC and the Haussmannian Paris. This new development is in fact located at the juncture of beautiful Haussmannian buildings aligned around Square des Batignolles and the new Parc Martin Luther King. Fully aware of the break in scale imposed on the new ZAC (joint development zone), one of the aims of this project was to ensure a gradual transition between these two centuries through a modulation of volumes in order to connect the two scales. Macro-objective of the project is to respond to the need for housing in the Île-de-France region and to provide services to the northwestern sector of the city. Another fundamental objective lies in environmental sustainability obtained through actions aimed at climate rebalancing and social redistribution, to which the building diversity corresponds. The eco-district, divided in lots 01 and 03, is a vast architectural and landscaping project organized around the rue Cardinet, the upper terrace, the new street and the web of railways. All the roofs of lot 01 are of the "roof-terrace" type, which is to say with planted gardens. The artificial topography created by the roofs of lot 03 enabled the creation of links between the buildings on the south side (student housing) and the north side (family housing). The green roofs are enhanced by the presence of a series of bleachers. Lot 01 contains commercial spaces, middle income rental apartments and first-time buyer apartments, as well as a residence hall for doctoral students. There is a 460-place parking garage at the basement of the building made available to the entire district. The project connects three distinct elements by playing with the contrast between a black monolith along the SNCF railways, with inward and outward folds of a white chrysalis, and ends with the angular bleachers of an anodized bolder, the "cannon barrel" (office building).

ZAC Clichy Batignolles.
Authors: Avenier Cornejo Architectes + Gausa Raveau.

INFRA/PUBLIC

RESEARCH

#requalification #urban green #eco-district #mixité

# #ISP02 | PROMENADE PLANTÉE

Paris, FRA
status | 1986 - 1993
typology | PUBLIC PROJECT

The Promenade Plantée René-Dumont, also called Coulée Verte, is a linear green space used as a pedestrian promenade and public park located in the 12th arrondissement of Paris. The entrance to the Plantée promenade is located at 44 rue de Lyon: there are several gardens that develop along the route which has an area of 3.7 hectares and a length of 4.5 km. The walk, raised above the street level of the boulevard that runs along it, begins near the Opera Bastille, at the so-called Viaduc des Arts and ends near the boulevard that acts as a ring road to the Montempoivre gate, at the intersection of the Boulevard Carnot, Avenue Émile Laurent and Rue Édouard Lartet).

There are other examples of disused railway lines converted into a park or a promenade, however, the Promenade Plantée is the progenitor of the raised green spaces that winds through the urban fabric using a single viaduct. Designed by landscape architect Jacques Vergely and architect Philippe Mathieux, the Promenade plantée takes advantage of the old line that since 1859 connected the gare de la Bastille to Verneuil-l'Étang via Vincennes. Deactivated on 14 December 1969, the line was partially integrated into Line A Île-de-France of the RER (Réseau express régional), leaving the section Paris-Vincennes abandoned. The area was renovated starting in 1980. In 1984 Bastille station was demolished to make room for the construction of the Opéra Bastille.

The ZAC (Zone d'aménagement concerté) Reuilly project started two years later, in 1986. This project involved the green space recovery of the railway line between avenue Daumesnil and Montgallet and Reuilly streets, including the stretch of Promenade plantée between the place de la Bastille and the Porte de Montempoivre.

The path between bridges, tunnels, viaducts, highlights the gutting operation of the railway which is transformed into a continuous green canal. The promenade is divided into two sections: the first, the Viaduc des Art, is represented by a long hanging garden 9 meters wide and made up of a plant heritage of almost 200 different species, the second, the Promenade Verte, is a straight stretch under the road plan, characterized by the recovery of a railway tunnel in which spontaneous and wild vegetation has been maintained and integrated with new plantations. From a botanical point of view, it is interesting to observe that in many points of the walk the spontaneous plants born during the period of abandonment have been deliberately preserved.

# #ISP03 | ABANDOIBARRA

Bilbao, SPA
status | 1992 - 1999 (MASTERPLAN) - 2015
typology | COMPETITION - PUBLIC PROJECT

Within the complex transformation process that the city of Bilbao went through at the end of the twentieth century, the project of environmental recovery and regeneration of the Abandoibarra railway yard is part of a strategic plan that involves 5 different areas, including Zorrozaurre and the former freight station of Ametzola, which is complementary, both from an economic and functional point of view, to the area of Abandoibarra. The period of severe crisis in the steel industry has become an opportunity for the regeneration and urban development of the city of Bilbao. In 1987 the Plan de Revitalizaciòn was implemented, including the redevelopment of the Abandoibarra airport. Key moment for the revival of the city is the design of the Guggenheim Museum by Frank O. Gehry between 1993 and 1997. The museum immediately becomes an urban magnet, bringing the city to the international limelight, and producing chain effects, including the increase in the values of the neighboring commercial areas and the recovery of the real estate market. The masterplan provides for the relocation of industrial areas, the insertion of over 120,000 square meters of public areas between squares and linear parks, the enhancement of the river banks through the addition of pedestrian paths, as well as a direct connection with the University of Duesto which is extended by creating a direct axis with the city center. The tram connection with the central station is also strengthened. In addition to the objective of giving identity to an area that was once a railway and industrial area, the project aims to transform the district into an attractive pole thanks to the presence of the Guggenheim museum and the opera house: the aim of the masterplan is precisely to direct connection between the two buildings. Another fundamental principle was the development of a direct connection with the motorway network to enhance and facilitate the aspects of mobility. While abandoning the original mobility, the road and pedestrian system has nonetheless been increased and a new tram network has also been added to facilitate connection with the rest of the city. The public areas occupy about one third of the project and are available in many forms. We move from the long rows that draw the tramway, to the almost entirely plant-based parterres, to the hybridization elements between the vegetable and mineral elements, the main example of what the Euskadi square is. Another main theme of the masterplan is the redevelopment of the banks within the Nerviòn river, on this matter the Ribera park was created, a linear park that develops on two levels and extends for about 3 km to which playful cultural and commercial functions are attributed. The winning project was curated by Balmori Associates studio and RTN Architect studio, authors of the masterplan between 1992 and 1999. The huge urban transformations were carried out and implemented by Bilbao Rìa 2000, a public company founded in 1992 by the consortium of institutions regional and state owners of the intervention areas (railway and industrial complexes).

Abandoibarra aerial view, 2012.
Source: balmori.com/portfolio/abandoibarra-masterplan

INFRA/PUBLIC

RESEARCH

#mobility #identity #post-industrial #attraction pole

# #ISP04|KALVEBOD WAVES

Copenhagen, DEN
status | 2008 - 2015
typology | COMPETITION - PUBLIC PROJECT

Kalvebod Brygge is a waterfront area in the Vesterbro district of Copenhagen. The southern part of the area is to the west bounded by an extensive railway area, a section of which is now under redevelopment into a linear park with scattered buildings and a super bikeway, which will ultimately provide a greenway between the city centre and the South Harbour.
Kalvebod Brygge is situated opposite to the popular Copenhagen summer hang out, Islands Brygge. Kalvebod Brygge has the potential to be Islands Brygge's biggest urban counterpart but has, until now, been synonymous with a desolated office address devoid of life and public activities. The new waterfront, realized by Julien de Smedt Architects (former founder of the architectural practice PLOT in 2001), will be a place for a larger spectrum of public activities. With a close connection to the central train station and Tivoli, Copenhagen's famous city amusement park, 'Kalvebod Bølge', the 'Kalvebod Waves' will become a hub, buzzing with activities and providing a chance for the inner city to regain its connection to the harbour. Constituted more by its functionality than its tradition, this inner city site is less fragile than others and manifests Copenhagen's contemporary urban waterfront with neighbouring entities such as Black Diamond Library and Nykredit building. The project consists of two main plazas, which extend across the water and are positioned considering sunlight and wind conditions. To the south, the pier allows for a flexible public space on the water with facilities to host events related to the creative industry. During the last 10 years Copenhagen has developed into a stronghold for the creative class, therefore Kalvebod Brygge proposes an urban showcase that gives organizations, companies, festivals and fairs a location along the waterfront.
In connection with this space, an active water enclave is created, for various water related activities. The plaza and surrounding pontoons provide the necessary facilities for these activities to function. The flow of boats that commute to and from the water hub also creates an active maritime background and secures the connectivity of the plaza to the rest of the city. The second square acts as an oasis on the water, providing both proximity and access. This recreational space, with a beach, allows a break from the hectic pace of urban life, where a floating garden is proposed. A maritime park where urban and maritime life meet.
A multipurpose pier with a series of undulating bridges and promenades that rise from the water like waves, made of corten steel. Raised above the water on stilts, two concrete piers offer amenities including a dock for boats, a canoe club and an event space, while the covered areas with benches encourage sunbathing.

Kalvebod Brygge, Copenhagen.
Source: jdsa.eu/kal/

INFRA/PUBLIC

RESEARCH

#waterfront #footbridge #linearpark #swimming #bath

# #ISP05 | EUROPAVIERTEL

**Frankfurt, GER**
**status | 1997-2013 (MASTERPLAN); 2021 (COMPLETED)**
**typology | PUBLIC PROJECT**

Europaviertel is the last major inner-city development area in Frankfurt, centrally located between Messe Frankfurt and the main train station. A new downtown district that is being built on the site of the former main freight station in the Gallus district. Development work on the conversion area started in 2005 and the first building was opened in 2006. Upon completion, the urban development project should be built including offices, hotels, apartments, a school and social infrastructure, parks as well as shopping and leisure opportunities. The completion of a subway connection is only planned for 2025. Around 30,000 people will work and 8,000 to 10,000 people will live in the entire Europaviertel. This ratio could shift in favor of the number of residents due to the increased demand for apartments since around 2012. During the nineties, the railway yard, mainly intended for the transport of goods, became so obsolete and unsuitable to meet the demands that it forced the traffic lines to be closed. In 1999, Albert Speer & Partner (on behalf of Deutsche Bahn, the German railway) drew up a general urban plan for the development and management of the airport, providing for the inclusion of residential areas and green spaces. Based on the two property owners, the area is subdivided into Europaviertel West and Europaviertel Ost for a total of 90 hectares; the easily recognizable border forms the high-altitude railway line with the Emser bridge. It is a residential and commercial district: the construction of offices, apartments, hotels, green areas, recreational facilities, a shopping center and a conference center is planned. In 2020, with the completion of the construction and infrastructure works, the intervention led to the creation of a new urban area that hosts 10,000 residents and 30,000 workers. The project provides for an increase in public transport by 2022 with direct connections to the city center, to the central station of Frankfurt am Main, close to the airport. The extension of the railway network is also planned with subway lines that can speed up travel within the city. Overall, the redevelopment aims to create a part of the city with a set of functions capable of satisfying demographic development, acting as an union between the areas north and south of the airport. The east-west connections are located below the main park around which the buildings for tertiary use are concentrated. The main traffic artery culminates in the Europagarten, a park that offers a wide range of public activities. Next to the park there is a large green residential area. Adequate space will be provided for the long-term expansion of the trade fair and exhibition complex within its important city center. After the failure of the project of the Millenium Tower or T365, a skyscraper with an originally planned height of 369 meters, in September 2020 CA Immo Deutschland announced that it had signed an urban development contract with the City of Frankfurt am Main for two 260 and 150 meter high buildings next to the Skyline Plaza.

Europaviertel West, Frankfurt am Main.
Source: Dirk Laubner, schuessler-plan.de/en/projects/europaviertel-west.html

INFRA/PUBLIC

111 RESEARCH

#infrastructure #public spaces #europagarten #residences

# #ISP06 | SEOULLO 7017 SKY GARDEN

Seoul, KOR
status | 2015 (COMPETITION); 2017 (COMPLETED)
typology | PUBLIC PROJECT

Located in the heart of Seoul, a true plant village has been realised by Dutch architects MVRDV on a former inner city highway in an ever-changing urban area accommodating the biggest variety of Korean species of plants and transforming it into a public 983-metre long park gathering 50 families of plants including trees, shrubs and flowers displayed in 645 tree pots, collecting around 228 species and sub-species.
In total, the park will include 24,000 plants (trees, shrubs and flowers) that are newly planted, many of which will grow to their final heights in the next decade.
In the 1960s, a decade after the Korean War, Seoul planners ordered the construction of dozens of elevated highways to keep traffic flowing through the capital. A few decades later, the huge overpasses have become obsolete, often underutilized and risky infrastructures in terms of accessibility and safety.
Seoullo, the Korean name for Skygarden translates to 'towards Seoul' and 'Seoul Street', while 7017 marks the overpass' construction year of 1970, and its new function as a public walkway in 2017. The pedestrianised viaduct next to Seoul's main station is the next step towards making the city and especially the central station district, greener, friendlier and more attractive, whilst connecting all patches of green in a wider area.
MVRDV addressed the need to transform the forgotten and existing infrastructure into a green and sustainable symbol that will become a catalyst for a greener neighborhood for Seoul. The intervention also promotes the idea of a park as a cultural device that includes activities that involve the city. Iconic also for the context in which it is inserted and for the overlap between pedestrian public space and vehicular infrastructures on the street level. Together with the municipality, local NGO's, landscape teams and city advisers are committed to accommodating the biggest diversity of flora into a strictly urban condition. New bridges and stairs connect the viaduct with hotels, shops and gardens. The project also includes new satellite gardens that can be connected to the Skygarden, as additional ramifications already structure. With an approach of addition, recreational, residential and small-scale business modules have been integrated, such as bookstores, shops, flower shops, in order to move commercial and daily life flows to the top.

INFRA/PUBLIC

RESEARCH

Seoullo 7017 Skygarden. Author: MVRDV.
Source: mvrdv.nl/projects/208/seoullo-7017-skygarden

#platform #overpass #walk #public space #koreanflora

# #ISP07|THE HIGH LINE

New York, USA
status | 2000 - 2019 (COMPLETED)
typology | PUBLIC PROJECT

The project for a public park was born in the late nineties from the initiative of the civic organization "Friends of the High Line". Completed in 2019 and designed by Diller + Scofidio and Renfro in collaboration with James Corner Field Operations and Piet Oudolf, The High Line is a 2,5 km long public park built on an abandoned elevated railroad stretching from the Meatpacking District to the Hudson Rail Yards in Manhattan.
Inaugurated on June 3, 1934, the High Line was originally built to transport goods of various kinds. The intention was to create an entire manufacturing and storage district connected to the port area. The double-lane overhead viaduct constructed of steel and concrete pavement was capable of supporting two fully loaded freight trains. Following the strengthening of commercial air traffic, however, the rail freight sector was penalized and, in the 60s, the southern part of the structure was dismantled.
The last delivery of goods on the remaining northern part of the line took place in 1980. Inspired by the melancholic, unruly beauty of this post industrial ruin, where nature has reclaimed a once vital piece of urban infrastructure, the new park interprets its inheritance. It translates the biodiversity that took root after it fell into ruin in a string of site-specific urban micro-climates along the stretch of railway that include sunny, shady, wet, dry, windy, and sheltered spaces. Characterized by a succession of different urban microclimates, the railway section includes sunny and sheltered spaces.
Through a strategy of combining agriculture and architecture, the surface of the High Line is configured in paving units consisting of individual prefabricated concrete boards with open joints to encourage the emerging growth of spontaneous greenery. The long paving units create a structured landscape without preferential traces, where the public can move in an unscheduled way.
The park is home to the wild, the cultivated, the intimate and the social. The "Sperone", the last section of the High Line, has a large square for public events, rest and meeting areas. The High Line opened to the public in sections, starting in 2009, with phased openings in 2011, 2014, and 2019. From New York City's investment of $115 million USD, the High Line has stimulated over $5 billion USD in urban development and created 12,000 new jobs. Initially imagined as a singular, idiosyncratic, local solution, the High Line drew 8 million visitors and has "gone viral" as a global development model: over one hundred cities worldwide have been inspired to transform their obsolete urban infrastructure into public parks.

# #IIND01|HIGH SPEED TRAIN STATION

Logroño, SPA
status | 2006 - 2012 (COMPLETED)
typology | PUBLIC PROJECT

The project by architects Ábalos + Sentkiewicz for the construction of a new high-speed station is part of a more complex urban redevelopment vision which also includes landscape, ecological and economic themes. The plan will be complemented by a bus station, a park and a complex of residential buildings, five of which are tower blocks. The high-speed transport system is already very articulated within the country, with numerous lines that form an extensive system that can be divided into three passages between the Castejón station and the of Logroño one. In terms of spatial organization and functional program, the new railway hub is, in the words of Iñaki Ábalos, more like a contemporary airport than a traditional station.
The purpose of the high-speed line that includes the Logrono stop is to create a direct connection between the various Spanish stations, including those in Madrid, even connecting with France. Moreover, the station serves as a starting point for a new urban project that restores the connection between zones north and south in Logroño and at the same time serves to generate a great public park being the rooftop an integral part of this park, giving the geometric and topographic relevance to the volume. The faceted form of the roof corresponds with the park's one in the top, so both surfaces compose beam and underside in a folded plan which in the external side generates a hill or a lookout over the city and in the internal one, adopts a configuration similar to a cave or a grotto. These references to the picturesque tradition unify the whole project and its public dimension. The landscape component of the project is also significant, which is highlighted with the construction of the extensive park located on the roof of the station. Along the entire roof there are 14 skylights made with reflective steel mirrors that radiate light on the park during the night.
Of the entire complex of approximately 145,000 square meters, 8,000 square meters correspond to the railway station while the rest is divided between a new bus station, which has 10,800 square meters, parking and residential areas.
The underground nave is 400 meters long, 42 meters wide and 7 meters high. This space consists of two platforms and five platforms: four for public transport and one for freight. The tunnel measures a total of 1.4 km and the entire structure is supported by 18 pairs of pillars with dark gray finishes.

High Speed Train Station in Logroño.
Source: Wikimedia Creative Commons.

INFRA/INDUSTRY

117 RESEARCH

#connection #progress #high speed

# #IIND02 | STRATFORD CITY

London, UK
status | 2007 - 2020 (COMPLETED)
typology | PUBLIC PROJECT

The Stratford area is located east of London and, with Canning Town and the Royal Docks, forms the axis of urban development known as the Arc of Opportunity, the largest urban regeneration project in Europe today. For 300 years the area has been home to industrial sectors that have fostered London's economic growth thanks to the production of hydroelectric energy and plastics, taking advantage of the presence of the Lea River. In the 1970s, the closure of the nearby Docks and the gradual decline of the industry halted the growth of Lea Valley, leading to its degradation. Since 1997 Stratford has been a strategic hub for international transport thanks to the presence of the CTRL (Channel Tunnel Rail Link). The transformation project was promoted by the railway authority, which centralized the ownership of the site and strengthened its accessibility and connectivity. In 2005, the awarding of the London 2012 Olympic Games became an opportunity to initiate a significant regeneration process of the area and a catalyst for investments in new development projects, financed by both the public and private sectors, with the aim to renew and regenerate the infrastructures, public spaces, housing heritage, commercial opportunities and sports, cultural and educational facilities of the area. Works, begun in 2007 and led to the reconstruction of the disused industrial area along the banks of the River Lea. Among the proposed development principles it is relevant to mention the need to respond to the expected strong demographic growth, the elimination of the poverty gap between the existing neighborhoods and the rest of London, the social regeneration and neighborhood identity, the implementation of the services, finance, tourism and entertainment, the redevelopment and enhancement of waterways to promote biodiversity. Integration with the environment and the surrounding social fabric is the principle that has guided Stratford City's development strategies. To ensure the integration of the project into the context, the master plan created a flexible system for the development of high quality public spaces, incorporating a network of landscape connections, new roads, pedestrian areas and cycle paths, able to connect the communities close to the new urban development. Thanks to its proximity to the Olympic park, the station was used as the main arrival point for spectators of 2012 Olympics and Paralympics. Large equipped green areas have been designed and made available to citizens. The project focuses on the concept of environmental sustainability. The strength point of the program is the long-term vision of this development plan, which provides for the reuse of the area at the end of the Olympic Games to ensure that real estate investments bring a subsequent benefit to the whole community. The 73 ha site redevelopment project includes the construction of 6,000 new homes, for a total of 31,000 jobs and approximately 17 million passengers a year.

# #IIND03 | KING'S CROSS

London, UK
status | 2001 - 2020 (COMPLETED)
typology | PUBLIC PROJECT

King's Cross Central is one of the largest and most complex recently completed urban regeneration projects in Europe. The master plan for the 27-hectare site was an urban redevelopment process that integrated the development of diversified functions with the redevelopment of important industrial artifacts from the nineteenth-century period, similar to the Granary Building, already inaugurated in 2011. King's Cross is located in the London Borough of Camden, 4km north from Charing Cross and 4.5km northwest from Liverpool Street, within the London perimeter. The airport is characterized by a teardrop shape and is crossed horizontally by the Regent Canal which divides it in half. It is located adjacent to Euston Road and King's Cross and St. Pancras International train stations. Since the 1970s, industrial activities begin to enter a phase of decline and the structures are gradually being abandoned. The area is characterized by a strong negative impact which, since the 1980s, has affected the rental market, the lowest in central London, with a commercial stock unchanged since the nineteenth century. The turning point took place in 1996 with the decision to build a new connecting tunnel from St Pancras station to the Channel Tunnel: the Channel Tunnel Rail Link becomes an important incentive for the urban regeneration of King's Cross and the starting point for the relaunch of the entire district. The economic revitalization of the airport is the main purpose on which the entire redevelopment process is based through the search for an identity through urban design, accessibility to the entire sector and connection to the rest of the city, promotion of multifunctional and flexible projects, the use and enhancement of existing heritage, collective participation with a consideration on urban safety and sustainability. In addition to the coexistence of multiple types of services (commercial, office, residential, public and private), the intention of urban redevelopment also extends to the masterplan by defining surfaces with functions that can also be defined during its development. This expedient, namely the choice to define the surface in terms of "total permissible land" ensuring greater flexibility for approximately 20% of the GFA, made the project more adaptable to changes in the market, while leaving the routes and minimum heights fixed and maxima, the density and scale of the elements of the transformation. The project, completed in 2020, led to the construction of 50 new buildings and 10 new public parks and squares; in 2018, construction of the first Google headquarters outside the United States began at Pancras Square.

# #IIND04|ZUIDAS

Amsterdam, NED
status | 1998 - ongoing
typology | PUBLIC PROJECT

Strategically located between the Zuid and Buitenveldert areas, and close to Schiphol Airport, Zuidas (South Axis) is the financial center of the city of Amsterdam and the country. It hosts around 800 Dutch and international companies as well as one of the most important universities of Amsterdam, La Vrije Universiteit. An urban district in constant change, for example with the arising of the European Medicines Agency and the related increase in life sciences and health sectors, Zuidas is also rapidly becoming a center for research and experimentation in the medical field. The presence of numerous companies (ABN AMRO, JPMorgan Chase, MUFG Bank, BinckBank, Ax Finance, Ebury, Deloitte, EY, etc.) required efficient forms of connection. In fact, in the summer of 2018, a new metro line was inaugurated that connects Zuidas with the center of Amsterdam and one of the largest infrastructure projects in the Netherlands was built, the Zuidasdok, i.e. the extension of the southern ring road A10 and the expansion of Amsterdam's Zuid station. This area, already dominated by large-scale infrastructure and extensively used space, will be transformed over the coming years into a metropolitan area of national and international stature. Particularly strategic is the aim to develop a large railway station transversely to the main infrastructure routes and to the newly designed built strips. To realize these aims, the infrastructure in the area will be channelled underground, following the so-called 'dock' model: motorway, metro and railway will be moved underground in phases. Furthermore, the capacity of the ring-road will be increased and the WTC station will become a high-speed railway station of European stature. The lowering phase of infrastructure underground will create space for the new centre totalling 800,000 m2 in area. Within this context priority is given to the creation of a mixed use urban environment, with a ratio of c. 45% dwellings, 45% office space and 10% amenities, in high density. The expansion of WTC station will create a major transport intersection where buses, trams, metros, trains and high-speed trains link up smoothly. Together with the motorway access route, this public-transport intersection guarantees perfect access to the new urban centre through all kinds of transport. The complete development of the Zuidas includes a program for at least 2 million square metres to be built in three phases and scheduled for completion by about 2025. Extending over 2.5 sq km of territory, Zuidas is conceived as a financial center. However, the most important sector is the one related to sciences and medicine, the EMA. The building was started in 2018 and completed in 2019, and can accommodate about 900 employees from around 30 countries. Another major intervention is 2Amsterdam which will breathe new life into two office blocks in the heart of Zuidas, known as the Twin Towers, which will be given a 24/7 purpose that matches the ambitions of Amsterdam Zuidas.

INFRA/INDUSTRY

#skyscraper #urban-hub #connections #diversity

# #IIND05|ROTTERDAM CENTRAL STATION

Rotterdam, NED
status | 2013 (COMPLETED)
typology | PUBLIC PROJECT

The aim of the Rotterdam Central Station reconstruction project is to integrate the new station into the urban fabric of the city centre, transforming it into one of the main hubs of the European transport network, thanks also to the recent development of the High Speed Line (HSL). One of the main challenges of the project concerns the different characteristics of the urban space to the north and south of the station. The northern side, De Provenierswijk, is a 19th century residential neighbourhood, which represents the "provincial" side of the Central Station: it is a rather compact portion of the built-up area that remained almost completely intact during the extensive bombing suffered by the city during the Second World War. The south area of the Central Station, on the other hand, is the side with the "metropolitan" identity, rebuilt since the 1950s and the site of great architectural experimentation and typological heterogeneity of the buildings. Entrusted with the project in 2004, Team CS, a collaboration between Benthem Crouwel Architekten, Meyer en Van Schooten Architecten and West 8, involved in the project for the architectural, urban and landscape aspects, worked on the original building, renovating it and extending the north-south connecting cycle tunnel. It also concentrates the development of the city towards the station area, in order to anchor the new hub to the urban fabric while enhancing the surrounding context.
In order to emphasise the dual nature of the two parts of the city in relation to each other, the architecture of the station shows two different characters depending on the fronts. The entrance on the north side has a modest design, appropriate to the character of the Provenierswijk district and the smaller number of passengers. It is gradually connected to the city and the presence of greenery is accentuated by the transparency of the long façade. In contrast, the large southern entrance is clearly the gateway to the city centre. Here the station derives its new international and metropolitan identity from the glass and wood lobby. The roof of the hall, completely clad in stainless steel, gives rise to the iconic character of the building and points to the heart of the city. In this way, the central station takes on the structure and dimensions of the urban landscape in which it is embedded, measuring itself against the heights that characterise the metropolis and at the same time reflecting the human scale. The esplanade in front of the station is a continuous public space. In order to achieve this simplicity, the car parking spaces (750 spaces) and bicycle parking spaces (5200 spaces) are located underground below the square. The tram station has been moved to the east side of the station in order to not obstruct the connection between the station and the city. Buses, trams, taxis and the short-term parking area are integrated into the existing urban fabric and do not constitute barriers. The red stone of the station floor continues into the forecourt, merging the station with the city.

# #IIND06|YANGPU RIVERSIDE

**Shanghai, CHI**
status | 2016 (INTERNATIONAL COMPETITION); 2018
typology | PUBLIC PROJECT

The riverbank management project of the Yangpu River, in its encounter with the majestic Yangpu Bridge, is based on an initial consideration of the purpose of urban renewal of the entire riverbank area based on innovation and culture through knowledge and reuse of its fantastic industrial heritage as well as the use of the new road networks approved and in execution. From this perspective, the first purpose of the spatial reorganization project is to take care of the rationalization and use of the detected programmatic areas: the riverside area, consisting of a public walkway and industrial heritage, which are transformed into cultural and leisure centers; the new layout of Yangpu Road, forming a continuous, consistent and mixed-use urban fabric that integrates residential, commercial and office uses, the central area between the river and Yangpu Rd., in which green areas are deployed and uses dedicated to new forms of work and innovation. This first scheme in three large strips allows a rationalization of uses, tracings and typologies giving rise to some approaches that try to create identity taking advantage of their morphological characteristics and generating with the set of industrial buildings a set as a collage that tries to make the most of existing and emphasize the interest and variety of the urban layout and especially the public space.

With the improvement for the functional positioning and development strategy of Yangpu Riverside, the urban design is to clarify the image orientation and key concept of Yangpu Riverside development based on the resource conditions of the century-old industrial civilization of Yangpu district. Overall urban design with the layout space includes industrial positioning and functional planning, development scale and spatial structure, waterfront open space and landscape features, infrastructure support and underground space utilization. Started with an international competition in 2016, the project is managed by the Shanghai local government on an area of 50,6 ha. The architectural and urban design, signed by architects Iñaki Ábalos, Renata Sentkiewicz (Abalos + Sentkiewicz AS +), celebrates the remnants of Shanghai's riverside industrial heritage. The layout reconnects the city with the coast by reusing the material language of the historic site in an innovative way. Structures, scratches and textures have been kept as the most vivid reflection in the history of the site, with the purpose of materialization of memory.

Through a low-impact design strategy, a system was designed to restore the ecological environment based on the spatial conditions of the factories and the local ecosystem. Beyond the original flood containment wall, an area of the valley rich in aquatic plants has been preserved through the inclusion of an ecological wetland that collects rainwater. The plants for the vegetation are mainly reeds and gazebos resistant to water and present in the local flora.

INFRA/INDUSTRY

#river #public space #basin #culture #memory

# #IIND07 | AUCKLAND WATERFRONT

Auckland, NZL
status | 2012 - (ONGOING)
typology | PUBLIC PROJECT

Growing from a population of 1.5 million, Auckland is expected to be home to 2.2-2.5 million by 2041. Auckland has a diverse ethnic and cultural composition and is currently home to over 150 ethnicities. Auckland's diversity is likely to continue to increase, and the Asian and Pacific proportion of the population is likely to grow most significantly. Younger, older and more diverse, this changing face of Auckland will need to be planned for and celebrated at the waterfront, which, in this way, constitutes a vivid interface between the sea, its uses and activities, and the inner city. According to the Waterfront Plan drafted in 2012, the waterfront is expected to be a major driver of Auckland's economic future. By 2040 the waterfront redevelopment will contribute $4.29 billion to Auckland's economy. Over the next 30 years Auckland's waterfront redevelopment will directly support 20,000 new full-time jobs and will contribute indirectly to a further 20,000 jobs across the region. The Waterfront Plan sets out the vision and goals for the waterfront and a range of short, medium and long-term initiatives to be delivered by Waterfront Auckland in partnership with a range of other parties including other landowners.

To achieve the Waterfront Vision, Waterfront Auckland has set five goals for Auckland's waterfront to be advanced by bold leadership. They are addressed especially to generate a blue-green waterfront, a resilient place where integrated systems and innovative approaches are taken to enhance the marine and natural ecosystems, preserve natural resources, minimise environmental impacts, reduce waste, build sustainably and respond to climate change. Others goals intend to generate a smart working, connected and liveable public space planned to achieve a significant lift in productivity, a place for authentic and gritty waterfront activities: the marine and fishing industries, water transport and port activities. A public waterfront will also provide free access to the water with new piers, walkways and bridges, pools and beaches. Waterfront Auckland aims to high-quality urban design and architecture choices. Wynyard Quarter has the capability of delivering a seamless extension to the city centre: it can accommodate growth in large numbers of office workers and residents alike. Outstanding design and architecture means: developing attractive spaces, a coherent design across the waterfront, places of surprise and interest, and delivering comfortable public spaces with weather protection, seating, sun, pedestrian priority and traffic-calmed streets that are safe and full of activity. The award-winning Urban Design Framework for the Wynyard Quarter was developed in 2007. Four key concepts of the Urban Design Framework for Wynyard Quarter are appropriate to the wider waterfront, such as: waterfront axis by establishing the 'waterfront spine', park axis by creating a landscape network, wharf axis by connecting land and sea, waterfront precincts by developing areas of distinct character.

Auckland Smart Green Waterfront Development.
Source: goexplorer.org/auckland-smart-green-waterfront-development/

#waterfront #auckland #reconversion #industry

# #IINF01 | SCALO PORTA ROMANA

Milan, ITA
status | 2020 - ONGOING
typology | COMPETITION

The lack of relations between the northern and southern urban fabric is the starting point for the discussion on the redevelopment of the former railway yard of Porta Romana. The northern part of the city is a consolidated and histornorthern part of the city ic urban reality, while the urban fabric in the southern part of the railway yard is identified as an area undergoing strong transformation. The insurmountable barrier represented by the railway station in past years is linked to the original industrial-artisanal and logistics characteristics, connected to the presence of the freight yard. Like the other disused railway yards in Milan, Porta Romana yard is the subject of an ambitious transformation project on a metropolitan scale. The opportunity arose when the International Olympic Committee decided to assign to Italy the task of organising the Milan-Cortina 2026 Winter Olympics. In this process, Porta Romana is the chosen site for the Olympic Village, becoming a major economic and social catalyst for the city. Following the International Competition for the Masterplan of Scalo di Porta Romana launched by the Municipality of Milan in 2020, the winning solution "Parco Romana" is signed by the consortium formed by Coima, Covivio and Prada groups. As provided for in the Programme Agreement on the railway yards, "Parco Romana" was the subject of a digital Public Consultation process to collect comments and observations from the citizens, on the basis of which the designers integrated the Masterplan. The total budget is 180 million euro and will implement a large real estate development in the area of about 190,000 sqm. The result of a collaboration between 70 firms led by US architects Skidmore Owings & Merrill, the village will occupy about 60,000 square metres of residential, office, social and student housing space. With "Parco Romana", Milan conceives a new type of integration between nature and the city for a better quality of life: on the one hand, it regenerates a railway axis, on the other it generates a community of residents, athletes and visitors around a large public park, which is connected to the surrounding area thanks to the public transport network. The heterogeneous composition of this development offers a high quality environment for residences and working spaces, in symbiosis with the landscape and outdoor activities. The mix of public spaces and the accessibility of routes make essential services and commerce close by. In order to root the project beyond the Olympic event, the Masterplan foresees the commissioning and handing over of the Athletes' Village area to the Olympic Committee well in advance of the Games. Temporary functions, including restaurants, sports areas, commercial units and other public event spaces, are integrated within the complex. In terms of infrastructure, the project resolves the north-south division. The still active railway line within the area will be partially buried and covered by a park with facilities such as children's playgrounds, sports areas and urban gardens.

Rendering of the masterplan of the Scalo di Porta Romana, Villaggo Olimpico 2026 project.
Source: scaloportaromana.com

INFRA/INFRA

#connection #public transport #urban barrier #event #railway

# #IINF02|SCALO LAMBRATE

Milan, ITA
status | 2020 - (ONGOING)
typology | COMPETITION

The Lambrate site is also part of the plan to regenerate 7 disused railway yards in Milan. Located in the eastern sector of the city about 500 metres from Milan Lambrate Station, the site is a marginal area of about 7 hectares, outside the railway belt that separates it from the Città Studi district, between the historic districts of Lambrate to the north and Ortica to the south. The presence of the railway-metro interchange makes this site highly accessible to the urban and metropolitan areas. The terminal is part of an urban context in which significant transformations have already taken place and represents an important opportunity for redesigning the areas perimeter around the railway and reconnecting the existing urban fabric in continuity with the existing functions and public spaces. Thanks to the coordination of FS Sistemi Urbani and the platform for the management and communication of international design competitions C40 Reinventing Cities, Lambrate depot has been the subject of a complex public consultation process that has led to the winning proposal designated in 2020, at the stage of expressions of interest. Phase 2 of the competition process closed in March 2021. The winning proposal "Lambrate Streaming", with Sant'Ilario Società Cooperativa Edilizia as representative team and Caputo Partnership International srl as main architects, proposes a system of three interconnected squares: the "central square" and the "garden squares", creating a sense of collective life. Over 80 per cent of the area is dedicated to public spaces and public use, through orchards, educational and community gardens, equipped recreational areas, playgrounds and sports areas.The railway wall will be reimagined as an "urban backdrop" enriched with poems and verses. The heart of Lambrate Streaming is its 41 500 m2 of public park space surrounding the buildings. The wooden landscapes and low carbon building materials make the project unique from a sustainability perspective. Lambrate Streaming promotes eco-friendly transport via the construction of 1.2km cycling and 2.1km pedestrian walkways respectively. Both the renovated and new buildings aim to reduce carbon emissions using a variety of technologies and building practices whilst at the same time being able to maintain the cultural identity of the community. Lambrate Streaming aims to deliver clean energy, sustainable water and waste management as well as technological improvements to help mitigate undesired inefficiencies and promote a green and mutually beneficial environment to the citizens of Lambrate. The winning project proposes the creation of a large public park of 41 500 mq and a plan to plant 900 trees. 87.02% of the total project area is destined for public use. Waste and building management systems ensure that 75% of construction waste is recycled and that low carbon materials are utilized. The plan includes over 19,000 mq of affordable housing. The energy and water systems are optimized, in fact Lambrate Streaming provides 100% renewable energy and reduces water usage by 40% via rainwater collection and efficient building infrastructure.

# #IINF03|LA ANTIGUA ESTACIÓN

Burgos, ESP
status | 2008 - 2016 (COMPLETED)
typology | PUBLIC PROJECT

Built in 1905 by the Caminos de Hierro del Norte de España company, the old Burgos railway station has been out of use since 2008 after a new railway bypass was opened to divert train traffic from the original route. The project to rehabilitate the building and the connecting spaces around it is by Contell-Martínez Arquitectos of Valencia. The first railway building dates back to 1860, until the increase in rail traffic implied the need for a new station. The central pavilion housed the atrium, the ticket offices, the offices of the Station Master and Deputy Master and the telegraph office, with workers' houses. The first, second and third class waiting rooms were located on the left-hand side, as was the post office, and on the right side were the baggage room, canteen and some offices reserved for train drivers, security guards or technicians. Within one building there were different functions that had to work together, the same principle was used by the architects for the restoration project. The project by Contell-Martínez Arquitectos aims to rehabilitate the Old Railway Station to accommodate a recreational and leisure programme. The intervention extends, on one side, over the Station Square (former Travelers Courtyard) as a pedestrian and relationship space and, on the other side, a boulevard that is today where the railways used to be. In order to contextualize the building, a new pergola is built adapting its shape to this new boulevard. It serves not only as a transition between the scale of the building and the green area but also it remains to the great iron and glass marquee that protected the railways and the platforms. The intervention proposed in the Travelers Building seeks to recover its essence, adapting the construction to the new intended use. This is achieved by reinterpreting in depth the spatial whole conception, as well as the physical and visual relationship between the parts that characterize it as an architectural piece. The building is structured along a linear axis and is divided into the different areas requested in the program: in the east wing, the children's area; in the west one the restaurant and cafe area; in the access space, at the old mezzanine level, the administration and on the first floor, the youth area. Finally, the three towers are joined through walkways with stairways at the ends. The core of vertical communications and services is strategically built in the central body, this allows the individual use of the different building areas without the need to duplicate communication elements or toilets, while allowing facilities concentration. In the access to the building from the Travelers Courtyard, the original space is recovered thanks to two canopies that penetrate in the building, which also serves as windbreaks. These define the access point and the original geometry by reinterpreting that element. Inside, loading walls are cut down on the ground floor to achieve a fluid space and to get visual continuity between the different bodies.

# #IINF04 | CENTRAL PASILA

Helsinki, FIN
status | 2005 - 2009 (COMPETITION, 1st PRIZE)
typology | PUBLIC PROJECT

Keski-Pasila is a vast area of 167 ha a few kilometres from the centre of Helsinki and is characterised by the presence of a large railway station and a high concentration of roads, so much so that it has been defined as the Valley of Infrastructures. The Pasila district, already designed to relieve building pressure on the historic centre in the plan drawn up by Saarinen in 1918 and developed into various stages, is ready to take on a new important role in the city's new territorial dimension. Since 2003, in fact, Keski-Pasila has been the focus of a number of projects and investments through which the City of Helsinki has decided to create a core of high urban density and a great functional mix (services, commerce, offices and housing). In 2004, Cino Zucchi Architetti won the competition for ideas announced by the municipality, after which it drew up a general master plan for the area of the Pasila railway yard given back to the city. On the basis of the master plan, a memorandum of understanding was signed in 2006 between the City of Helsinki and the State of Finland (Railways) for the development of the area. The "fabric of towers" proposed by the Detailed Plan around a high quality open space has the particularity of combining in each tower the function of offices on the lower floors and residential use on the upper floors. In terms of both the urban design and the buildings, the project addresses the issue of energy saving and environmental sustainability in the new city, with Buro Happold's contribution on the latest technical and material solutions on the subject. The project has received great approval from both the public and the administration, and represents a highly innovative contribution to the theme of "sustainable urbanism". By 2040, the site will be occupied by 500,000 sq of new housing and 1 million sq of commercial space. The project is characterised by great public accessibility, both because of the presence of the railway station and because of the construction of new access roads, starting with the transformation of the existing bridge which will be converted into a new public-pedestrian axis capable of connecting the two neighbourhoods. Central Pasila will become a modern mixed-use urban centre, and the new architecture is expected to play its role as a strategic infrastructural hub reconnecting the urban fabric. Pasila railway station is the second busiest after Helsinki's Central Station and occupies most of the city area. Its job is to alleviate traffic congestion at Central Station, which is why all trains stopping there, both long and short distance, also pass through Pasila Station, even though the two stations are only 3.5 km apart. Since 2015, the city of Pasila has been undergoing a major renovation and so has the station area, which was demolished in 2017 and replaced by the new station, which opened in 2019. The project for the new Pasila station is part of a wider transformation of the large infrastructure theme area and includes the reorganisation of terminals for urban traffic, thus increasing the connection to Helsinki airport.

# #IINF05 | EUROPAVIERTEL

Stuttgart, GER
status | 1990 - 2020 (COMPLETED)
typology | PUBLIC PROJECT

The Europaviertel is one of the most important inner-city developments in Germany. The central freight and marshalling yard for supplying downtown Stuttgart was located there until the 1980s. Most of the A1 sub-area of the Stuttgart 21 master plan is located in the Europaviertel.
This covers an area of 16 hectares and building plots with a total area of 90,883 mq.
The headquarters of Landesbank Baden-Württemberg and Sparda-Bank Baden-Württemberg were built on the site of the former general cargo station of Deutsche Bahn between 1990 and 1996. These areas were built before the urban development concept framework plan for Stuttgart 21 was drawn up. Architects were Brunnert Mory Osterwalder Vielmo (that was Hans-Georg Brunner, Hasso Mory, Wolfgang Osterwalder and Manfred Vielmo) after their competition success.
In December 2001, the city of Stuttgart acquired the areas around what was back then the main train station from Deutsche Bahn AG. In contrast to most of the areas around the main train station, the areas in "sub-area A1" could already be built on before the new main train station was built. With the final development of the Europaviertel, the new main train station has to be moved to the city center.
The projects already implemented in sub-area A1 are the office building of the Landesbank Baden-Württemberg with the City Tower (construction areas 1 to 3, started in spring 2002), Südleasing office building (construction site 14, completed 2011), City library on Mailänder Platz (construction site 10.1, opened in 2011), Pariser Höfe on Stockholmer Platz (construction site 13, started in 2010), Sparkassenakademie (construction site 11, completed 2014), the Milaneo shopping center (plots 6, 8 and 9, opened October 2014), Cloud No. 7 (construction site 7, completed 2017), Europe Plaza with shops, offices (including the headquarters of McKesson Europe) and restaurants (construction site 10.2, completed in 2017), Tower on Mailänder Platz (construction site 5, completion 2021), Hampton by Hilton and Premier Inn (construction site 15, completion 2021).
Among the most important buildings constructed in Europaviertel there are the new library that was opened in 2011 as the first public building on Mailänder Platz in the Europaviertel. In the immediate vicinity of the Türlenstrasse and Budapester Platz underground stations, this location is ideally accessible via local public transport. Again, the Milaneo connects inner-city rental apartments, offices, a hotel, retail, gastronomy and services. Construction of the Milaneo began in 2012. A residential and hotel project called Cloud No. 7 was started in the summer of 2013 on the corner of Heilbronnerstrasse and Wolframstrasse and was completed in 2017. The owner of the 80-million-euro project is the Swabian housing AG.

Europaviertel Stuttgart.
Source: Wikimedia Creative Commons.

# #IINF06 | HUDSON YARDS

New York, USA
status | 2012 - 2022 (ONGOING)
typology | PUBLIC/PRIVATE PROJECT

Hudson Yards is a new urban transformation district of approximately 11 hectares that includes a complex of buildings and public spaces intended to encourage development along the Hudson River near a railroad depot on Manhattan's west coast, about 1 km from the Lincoln Tunnel that connects Manhattan to New Jersey. This is the busiest rail yard in the country, where 30 tracks converge in four tracks before entering Penn Station. The site is also directly connected to the 7 underground line. The complex was originally intended for other developments, notably in the early 2000s to become the site of the West Side Stadium, as part of New York's bid for the 2012 Summer Olympics. Public officials and private investors began developing the new Hudson Yards plan after the failure of the West Side Stadium. Construction began in 2012 on 10 Hudson Yards and the first phase of the site opened in early 2019. Hudson Yards is the centre of New York City's rapidly changing West Side. A hinge site between Chelsea, Hell's Kitchen, Midtown and the Hudson River waterfront, it connects these neighbourhoods previously separated by the site no longer in use. Phase 1 featured the construction of the performing arts centre The Shed, a public plaza, the scalable sculpture Vessel, and three residential high-rises on Eleventh Avenue. The investors plan to complete Phase 2, the western portion of the development, above the tracks between Eleventh and Twelfth Avenues. Phase 2 will provide additional office and residential space: when completed, 13 of the 16 planned structures on the west side of Midtown South will sit on a platform built over the West Side Yard, a depot for Long Island Rail Road trains. Hudson Yards is directly connected to the High Line, the pedestrian linear park built on a disused elevated railway, intercepting part of the route and integrating with it. The urban development of Hudson Yards, sponsored as an opportunity for real estate and commercial investment, aims to enrich the city with new economic activity and new jobs. The luxury flats, the shopping centre and the park surrounded by tall buildings convey that the idea of the complex is of a place mainly used for economic and commercial activities. The public square and gardens of Hudson Yards represent a green area unlike any other in New York. It was designed on a working railway yard and boasts a highly advanced irrigation, drainage and temperature control system. In five acres, more than 28000 different plant species and over 200 trees can be seen. Certified as the first LEED Gold Neighbourhood Development in Manhattan, Hudson Yards includes on-site power generation and is also equipped with two cogeneration plants and rainwater harvesting for on-site reuse. Hudson Yards, by Related Companies and Oxford Properties Group, is scheduled for completion in 2022. The project has been called the largest private real estate development in US history, and one of the most complex construction projects ever built in New York. Most of the land and buildings are not publicly owned but privately used for office or retail space.

Hudson Yards, The Vessel.
Author: Heatherwick Studio.
Source: heatherwick.com/project/vessel/

INFRA/INFRA

#infrastructure #railroad #vessel #technology #river

# #IINF07 | CARACAS METRO CABLE

San Agustín, Caracas, VEN
status | 2007 - 2010 (COMPLETED)
typology | PUBLIC PROJECT

The Caracas Metrocable re-connects a poor hillside neighborhood with the city focusing on social infrastructure and community development. Located in the informal barrio of San Agustin, responds to both transportation and socio-economic issues. Residents now have access to the city's existing public transport system, facilitating their commute around the city. The project also includes community facilities in each of the five stations, turning the Metrocable into an "infrastructure for social change". The company Urban Think Tank, founded by architects Brillembourg and Klumpner, adopted in their preparatory work a bottom-up approach, consulting community volunteers and working with the neighborhood to best respond to local residents' needs. The particular location of the barrio, on a hill and the densely constructed settlements, made the Metrocable the best-suited transport solution for urban integration and social acceptance.
The San Agustin Metrocable project was first developed in reaction to plans at the national level to construct a highway. In order to initiate an alternative planning process, a symposium was organized with the help of volunteers in July 2003 in order to give voice to professional urbanists and architects. One of the immediate results was the establishment of a task force, supervised by the Urban Think Tank. Consisting mainly of representatives of San Agustin, this group rapidly laid the foundation of the future San Agustin Metrocable by identifying it as the most appropriate and sustainable transportation mode for the neighborhood. Important criteria were the adaptation to terrain, a minimal invasion into the existing urban fabric and high design flexibility, determining the Metrocable as the most sustainable option.
As part of a day-long community workshop, the concept was further tailored to the needs and wishes of its future users. The construction of the Metrocable began in 2007 and was completed three years later. Its realization reduced cost and time of transport in order to offer residents new physical as well as social mobility perspectives outside of the barrio. The project further saw a reduction in violence due to the gondolas acting as an "eye on the city". Similarly, the Metrocable stations were conceived as community hubs to respond to social needs through the construction of community-oriented facilities. They include for instance spaces for community gatherings, health care facilities, a gym and a government-sponsored supermarket. The aim was to create a network of social infrastructure, easily reachable by residents and to bring together people from previously disconnected areas. The plan also foresaw the construction of 40 new housing units in one of the Metrocable stations in order to compensate for the necessary demolition of some houses in the barrio.

Caracas Metro Cable in San Agustin neighborhood.
Source: Iwan Baan, *iwan.com/portfolio/caracas-metro-cable-urban-think-tank/*

INFRA/INFRA

RESEARCH

#infrastructure #metrocable #social change

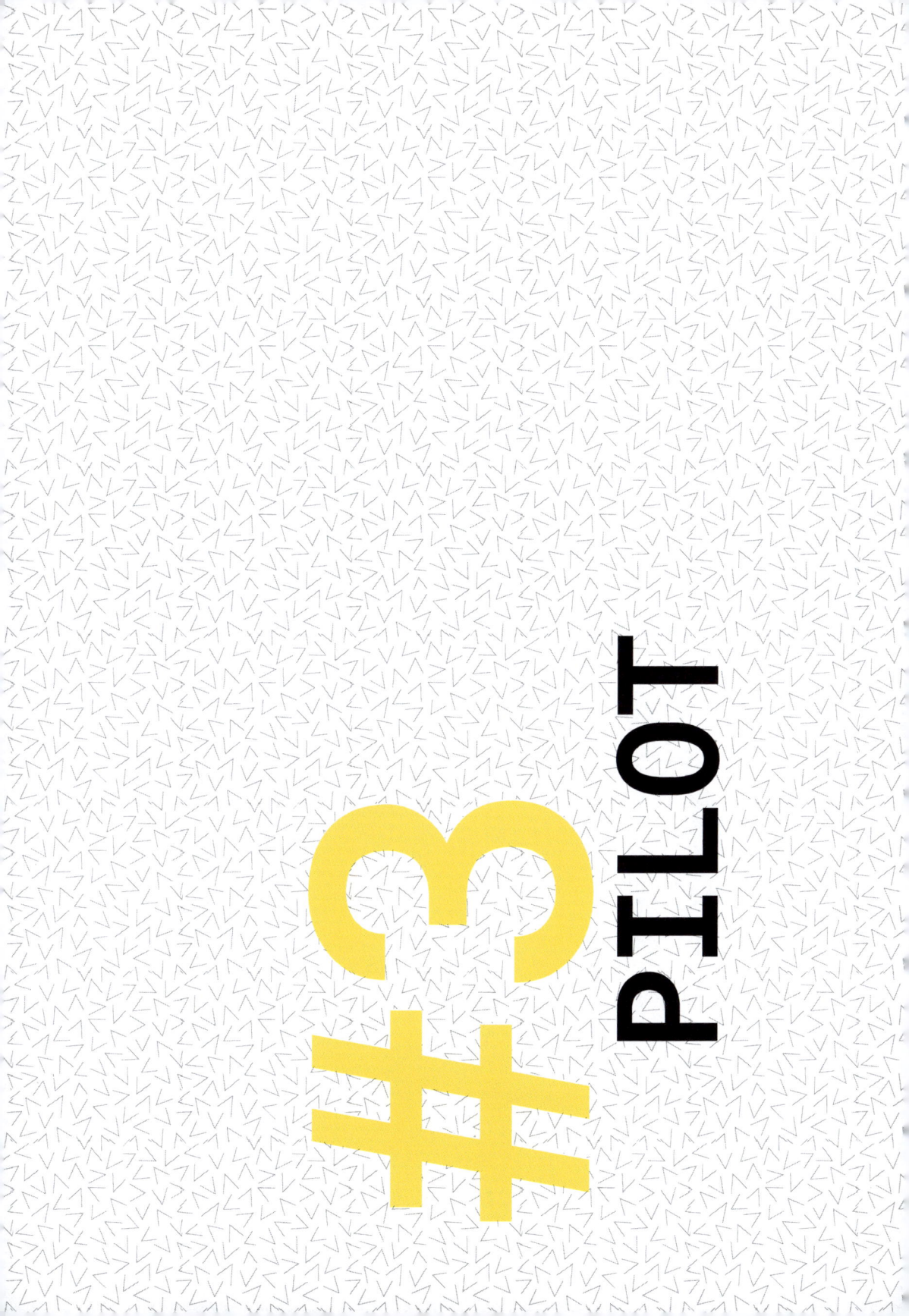

«The theme of sustainable urban regeneration is a central issue that must become a priority in the development policies of the coming years. For Italian architects, the urban issue is and will be the main problem of governments in these and coming years. In Italy as in the world, **the city and the habitat are at risk of "default" due to the depletion of energy resources and the poor condition of the building stock built after the war**».

- Consiglio Nazionale degli Architetti, Pianificatori, Paesaggisti - CNAPPC (2011)

# FRAMEWORK AND METHODOLOGY

**Roma Tuscolana**

Roma Tuscolana rail station is located in a semi-central position in the south-eastern quadrant of the historic city centre, in the Municipality of Rome VII, in the Tuscolano district of Rome VIII, within a compact urban context, characterised by a high population density, tertiary and commercial activities. The Tuscolano neighbourhood, the eighth of 35 Roman neighbourhoods, takes its name from Via Tuscolana that, unlike almost all radial roads in Rome - which form the boundary between adjacent neighbourhoods -, is located entirely within the neighbourhood. As to the origin of the name, the toponymic dictionary reads *strada che conduce al Tuscolo* (road leading to Tusculum). The boundary of the Tuscolano district runs along the Aurelian walls, from Porta San Giovanni to Porta Maggiore, follows Via Casilina up to Via di Centocelle as far as the junction with Via dell'Aeroporto, then follows it up to the junction with Via Tuscolana and continues along Via del Quadraro as far as Via Appia Nuova. At this point the boundary goes all the way up via Appia Nuova until it returns to Porta San Giovanni. The district is crossed by main urban routes: to the south-east by the crossing of via Tuscolana and via Appia Nuova, coinciding with the route of the Metro line A, and to the north-east by via Casalina Vecchia, which intersects a section of the Felice aqueduct, here resting on the ancient arches of the Roman Aqueduct of Emperor Claudius. The district is also crossed transversally by the suburban and metropolitan railway network, of which the Roma Tuscolana station is one of the most important modal interchange points for connections with the capital, such as Roma Termini and Roma Tiburtina.
This complex area has several criticalities such as noise pollution produced by active railway lines, which have not been provided with noise barriers in this section, and by vehicular traffic on Via Tuscolana and Via Appia Nuova. A further criticality to be considered is the formation of urban heat islands due mainly to high building densities and poor soil permeability with consequent air pollution. The redevelopment of the site of Tuscolana (46.300 mq) is one of the priorities of the Memorandum of Understanding between Roma Capitale, RFI and FS Sistemi Urbani for the "care of the railways" and the urban regeneration of disused railway areas, and is part of the *Green Ring* (Anello Verde) project, which envisages a continuous system of public spaces and facilities serving the city along the railway ring, between the Trastevere and Tiburtina railway stations. The Capital is part of the European project SMR-Smart Mature Resilience and has developed, as the first city in Italy, its own resilience strategy within the international project "100 Resilience Cities" that provides an innovative approach for the city by taking up several challenges that it will have to support in terms of urban resilience. Within this context and in line with the sustainability policies of the Capital City, in 2019 the global competition platform

*C40 - Reinventig Cities* brought together the owners of the Roma Tuscolana site (Ferrovie dello Stato Italiane, Rete Ferroviaria Italiana, FS Sistemi Urbani and Roma Capitale) in a collective urban and infrastructural redevelopment process. A competition of ideas was launched which ended in 2021 with the winning project "Campo Urbano" (Team representative: FRESIA RE SpA). At the heart of the competition, adopted as the methodological-strategic basis for the pilot projects in this book, is the pursing of specific objectives to improve the environmental quality of the site through targeted actions, such as pollutant remediation and acoustic and visual mitigation works of the railway infrastructure impact. The competition proposed that regeneration should take place through a process of "contamination" of spaces and functions and of integration of the urban system as a whole, avoiding "island" interventions not related to the neighbouring spaces. The projects should represent a model of urban development with characteristics of sustainability, resilience and experimentation and innovative architectural, typological and usage solutions.

**Methodology and tools for the pilot projects**

In line with the objectives, a set of pilot projects proposed for the redevelopment of the Tuscolana Station area were carried out. The methodology used is based on a three-stage process: (1) **Analysis of the context**: in which the variables of the scenario-framework in which the project action is carried out are identified and critically analysed: environmental, social, economic, political, technological, territorial factors, etc.; (2) **Definition of the concept**: in which the strategic approach to the project is outlined, the fundamental elements (actions and interventions) are defined and the objectives of the project proposal are set (functional schemes); (3) **Development of a vision**: in which a projection of the future scenario resulting from the project process is indicated, a synthesis of the transformations envisaged by the project.

In order to initiate the design process, an first cognitive survey of the Roma Tuscolana site was carried out through a series of **Interpretative Maps**, i.e. tools for exploration and communication. Each map focuses on a single system of morphological elements, providing a general analysis of the context of action, specifically:

1. SOIL (elevations, morphology)
2. BUILT (build and voids, urban fabric)
3. RESIDENCES (heights, typology, age)
4. PUBLIC SERVICES (education, transport, commercial facilities)
5. INFRASTRUCTURES (crossings, subways, bridges)
6. GREEN (agriculture, gardens and parks)
7. HERITAGE (historical artefacts, archaeology, emergencies)
8. PUBLIC SPACES (squares, car parks, open spaces, walkways)
9. BORDERS AND INTERFACES (linear, spatial, three-dimensional)

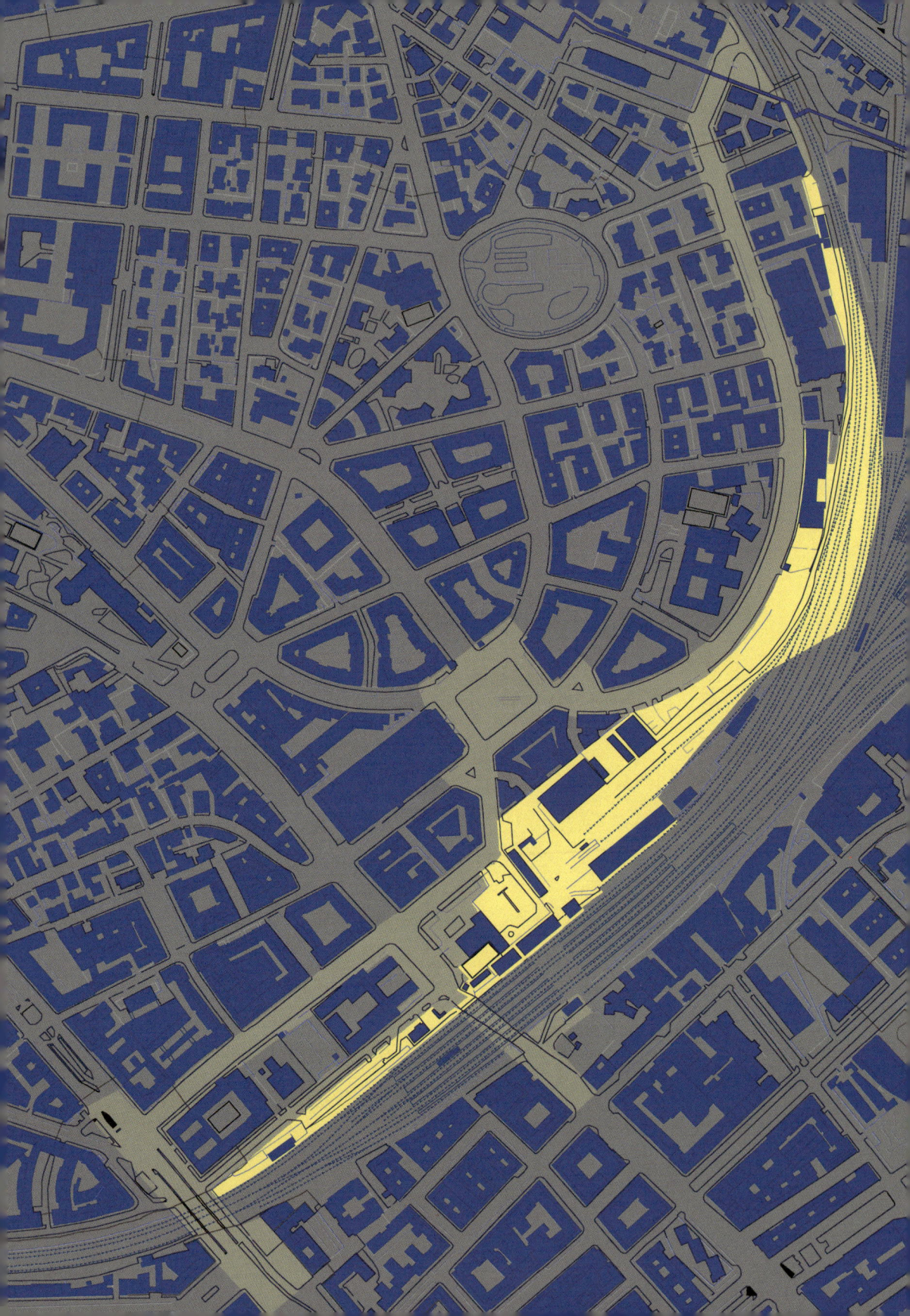

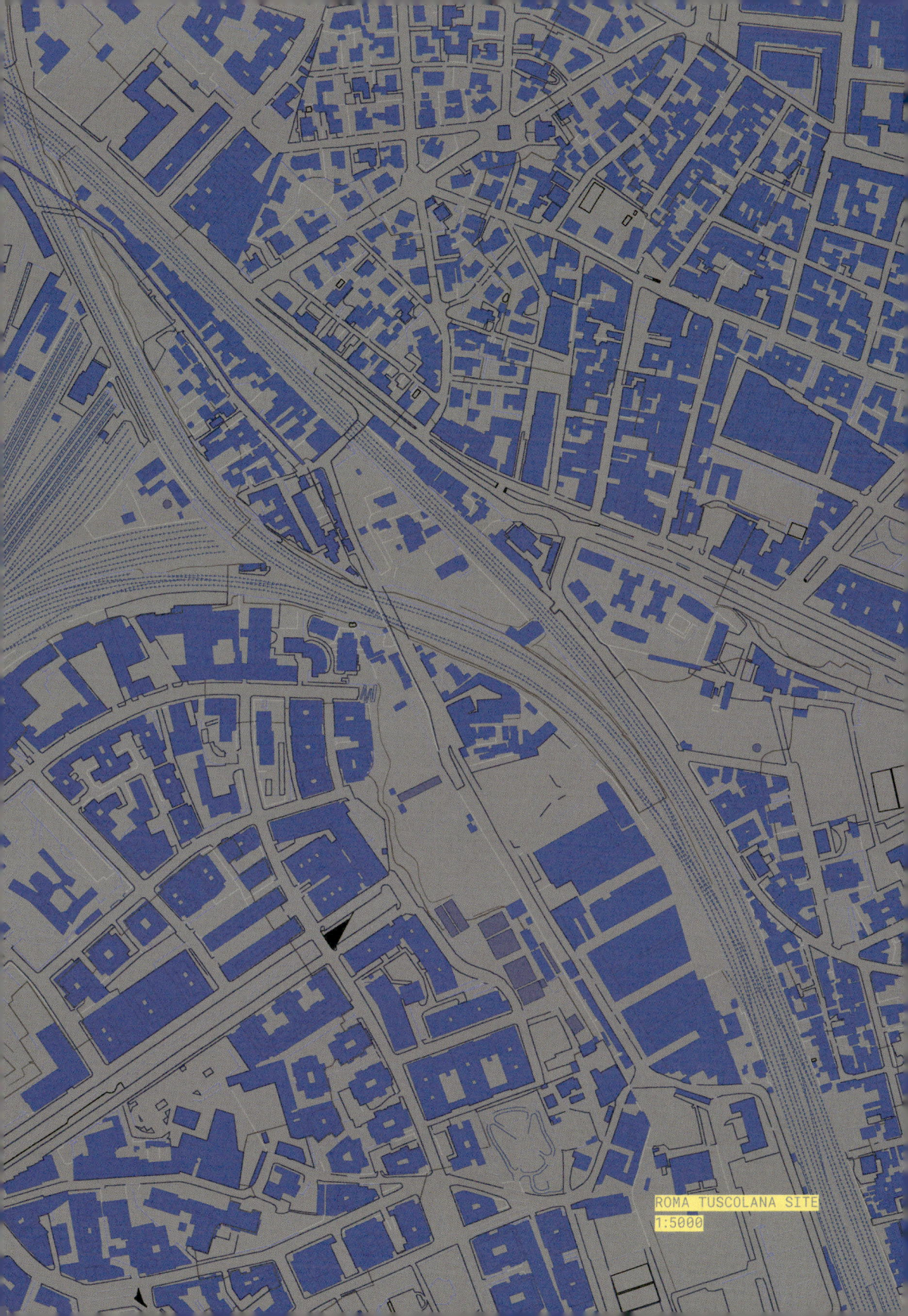

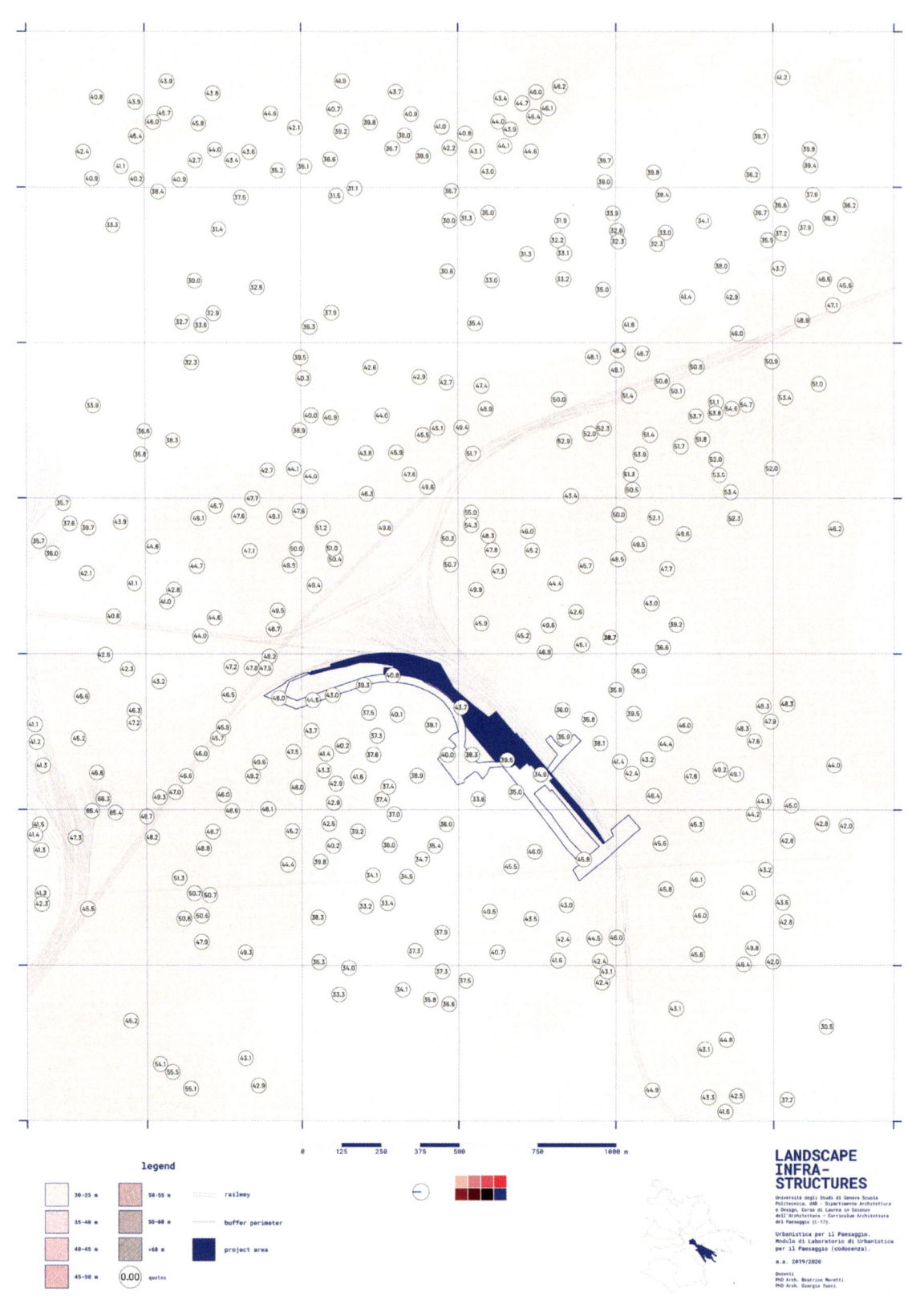

The soil map summarises the morphological
structure of the area of interest of Roma
Tuscolana station and the surrounding context.
It is interesting to note the strong difference
in elevation between the upper portion of the
station, above 40 mt in elevation, and the
lower portion, below 40 mt. Around the project
area, the surfaces in the immediate vicinity
of the station range between 40 and 52 mt in
altitude for the north-eastern portion, while
in the south-western quadrant the surfaces are
much lower, with a concentrated range between
30 and 40 mt.
From a water point of view, there are no
watercourses in the area; the nearest surface
water body is the Almone River, a tributary of
the Tiber, which is located to the southeast,
at a radius of approximately 2 km from Scalo
Tuscolano. On the other hand, there is an
underground water body at the limits of the
area, which crosses the Tuscolano district from
east to west.

I | SOIL

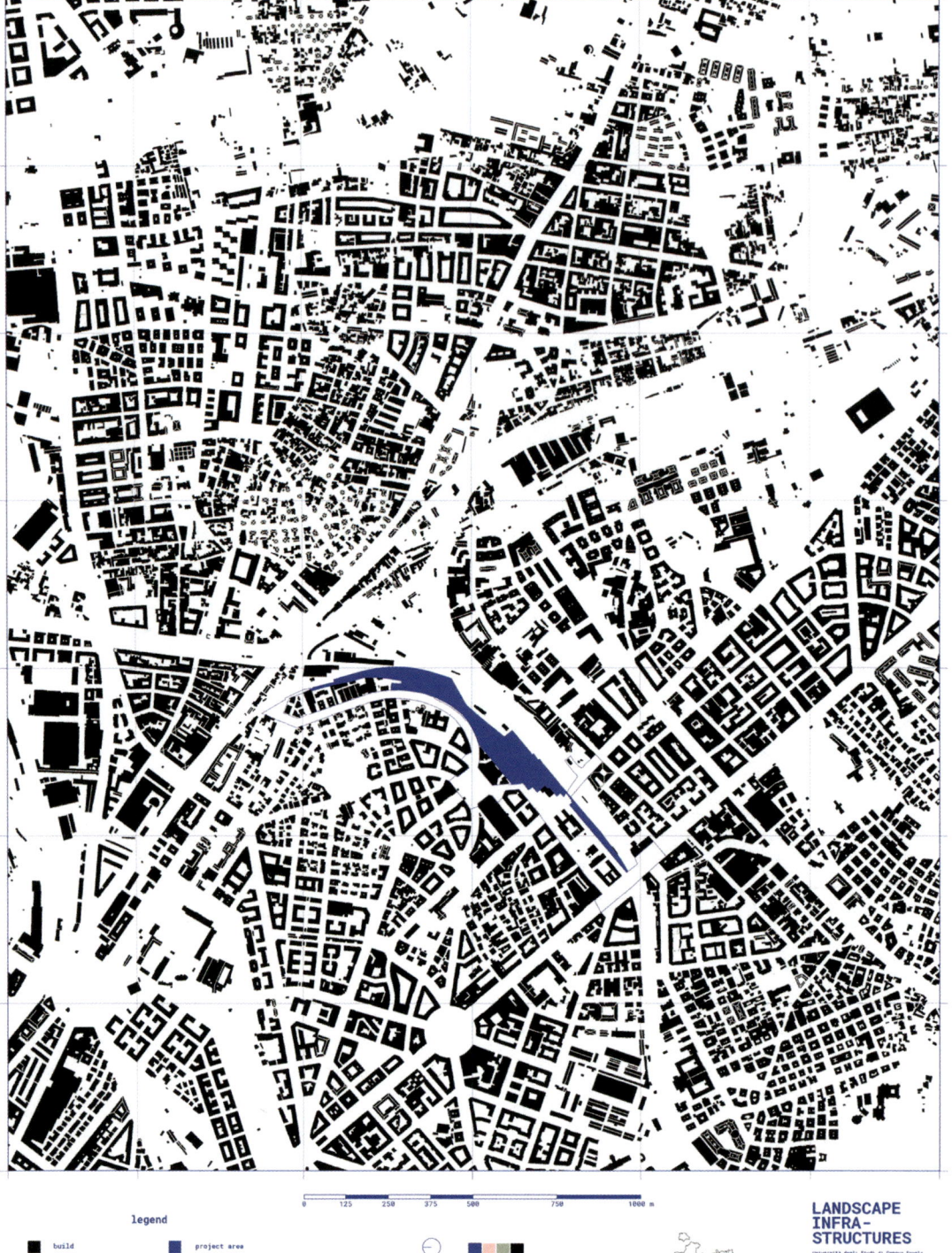

The map of the built environment highlights the volume of the urban fabric around the station. Starting at the beginning of the 20th century, the Tuscolano district was in fact the focus of strong urbanisation due to enormous building speculation that proved to be chaotic and wild. In 1921, a municipal measure led to the establishment of a number of important industries in the area, such as FATME, a telephone manufacturer, Sigma, a rubber manufacturer, Motor, a machine manufacturer, Pirelli and Montecatini, a chemical manufacturer. The map also shows the residential blocks built with the real estate development of 1949 when the INA-Casa institution was set up, which built the 'INA-Casa quarters' in the Tuscolano district for a total of over 112 buildings. Due to the high urban densification, the station area is totally occupied by non-permeable surfaces, most of which are made up of the railway network and the interchange areas of the Tuscolana Station. These are therefore spaces affected by buildings and their appurtenances, separated from each other by the railway and urban road network.

II|BUILT

# III | RESIDENCES

The town plan of 1931 increased the population density and brought the building limit to the Cinecittà plants that were to be built before the last war. Since then, with the increase in population density, the neighbourhood and the whole area underwent rapid development. The post-war period were the years of unrestrained urbanisation, speculative building and intensive ten-storey buildings prevailed in the district. Due to increasing immigration, the area was populated by immigrants from all over Italy looking for work and it was necessary to densify the residential structures to cope with the growing population. The project for the Tuscolano I, II and III neighbourhoods, developed by INA-Casa from 1949 onwards in the area between Via Tuscolana and the Parco degli Acquedotti - which took the name "Cecafumo" because of the fires from the shacks and artisan workshops that occurred within this border area, which had no outlet and therefore smoked the surrounding air - was highly significant. Today, with over 100,000 inhabitants, the Appio-Tuscolano district is one of the most densely populated in the capital. The map summarises the main residential typologies of the area divided into: single-family villas, residences (0-4 floors), residences (4-8 floors), residences (8+ floors), abandoned residences.

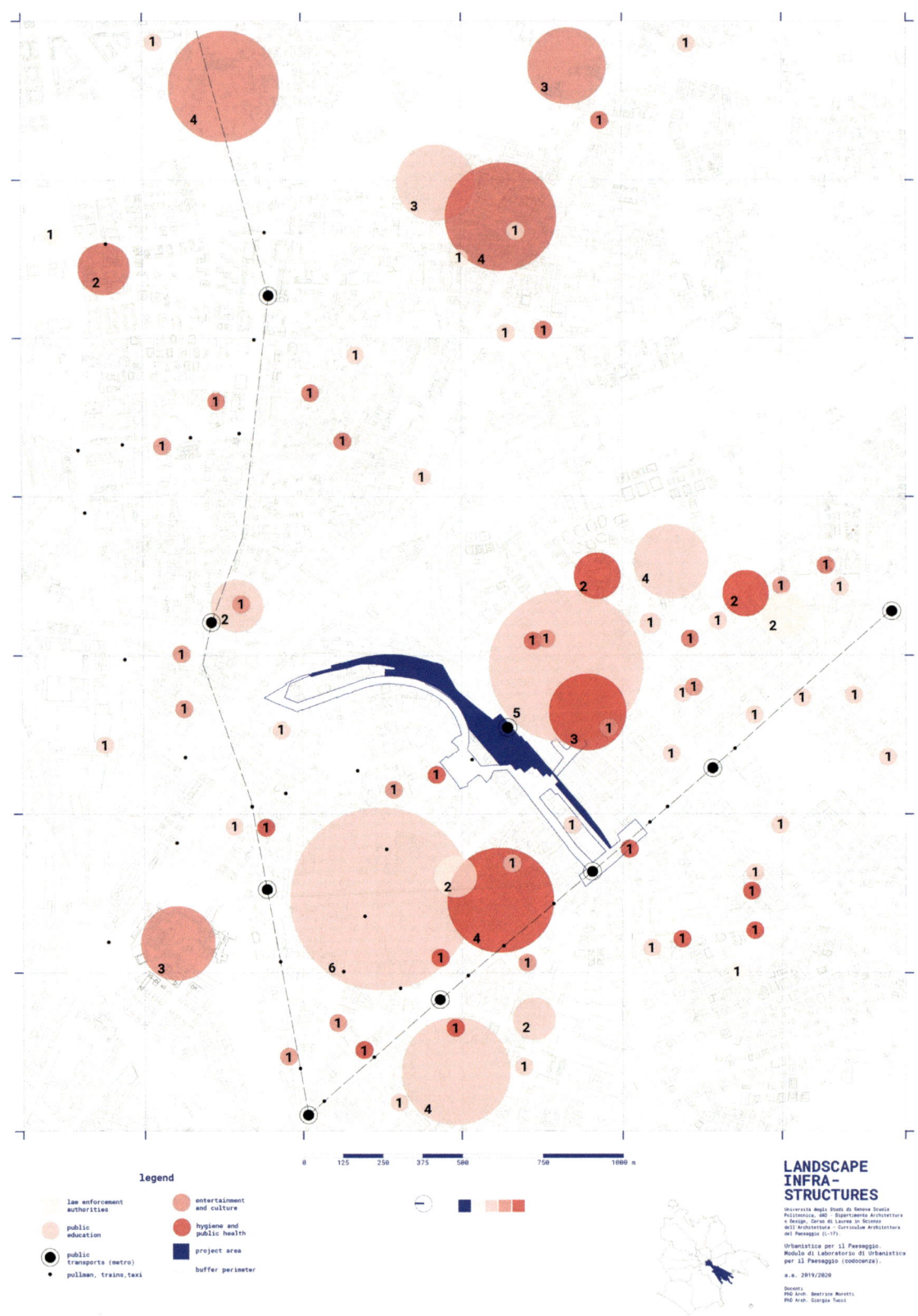

## IV | PUBLIC SERVICES

On the map there are several sectors of economic activity that are mainly located along the main roads, public transport services and educational institutions of various kinds. Particularly in the stretch from Porta Furba to Subaugusta, the district becomes a commercial hub, with a wide selection of independent shops and international chains, clothing and accessories shops, bars and restaurants.

Thanks to the station and the metro, the area is well connected to the city centre and, being a densely populated neighbourhood, is provided with numerous facilities such as schools, churches, theatres, monuments and archaeological sites.

Sub-area B (the lower part of the station) is characterised by the presence of commercial and sporting activities which represent an opportunity for the morphological and functional redefinition of the relationship between the station and the surrounding context, while the high schools in Via Gela and Via Adria represent an important reference for the provision of public services.

However, the nearby facilities such as green spaces, libraries, health centres, sports facilities, meeting places, etc. are still not sufficiently adequate.

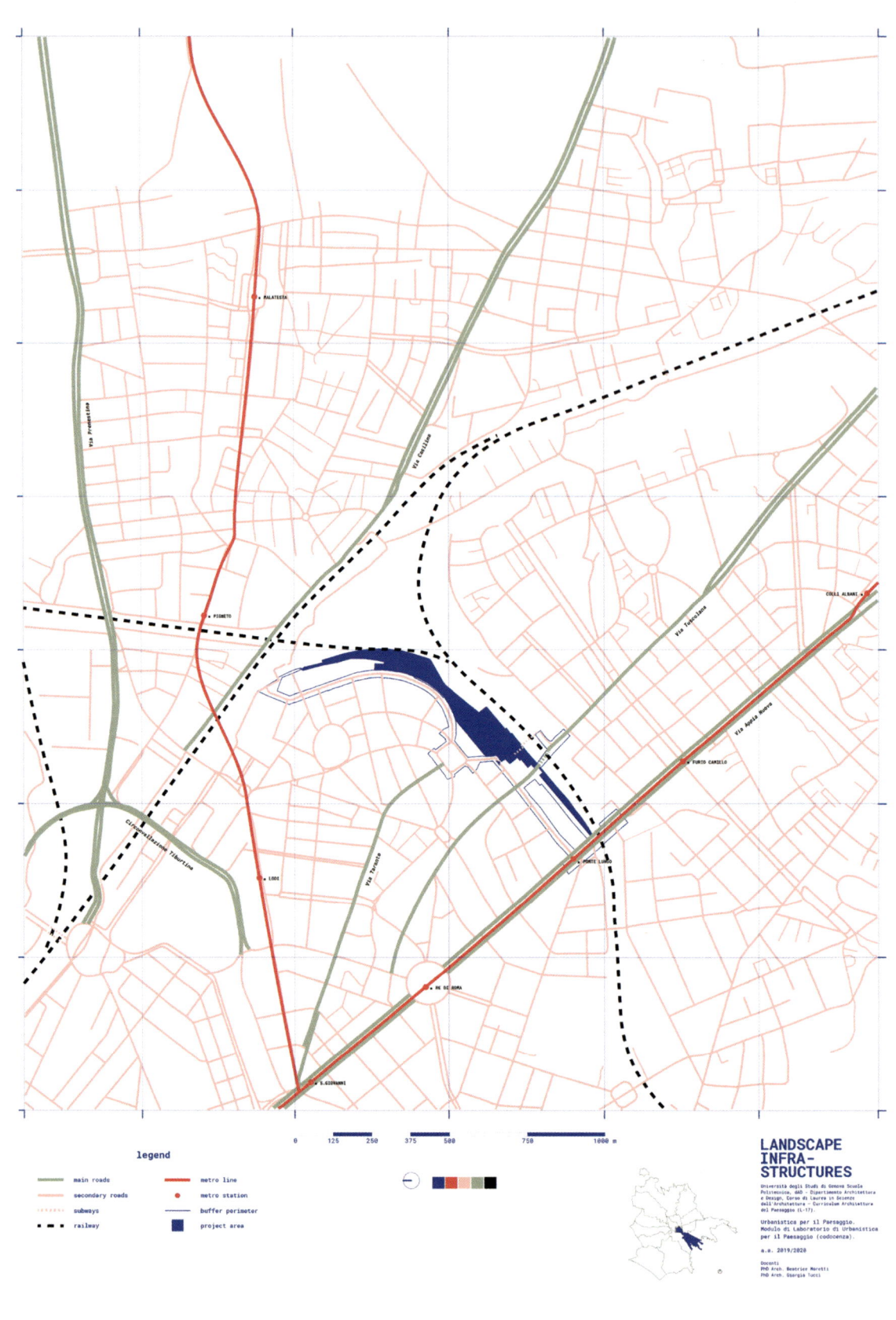

Within the urban context, the area is located in a semi-central position in the south-eastern quadrant aligned with the historic city centre and is developed close to the railway line. The area is crossed by one of the main railway infrastructure access nodes (south-east front) connected to the intermodal hubs of the Capital, such as Roma Termini and Roma Tiburtina. In particular, it is crossed by the FL1 Orte-Fiumicino Airport, FL3 Viterbo Porta Fiorentina - Roma Tiburtina and FL5 Roma Termini - Grosseto lines. Moreover, the Tuscolana station plays an important role in the local and city mobility system, also due to its proximity to the Ponte Lungo Metro A station. The map clearly shows all the main road infrastructures affecting the area: main and secondary roads, rail transport and the metro line.

V | INFRASTUCTURES

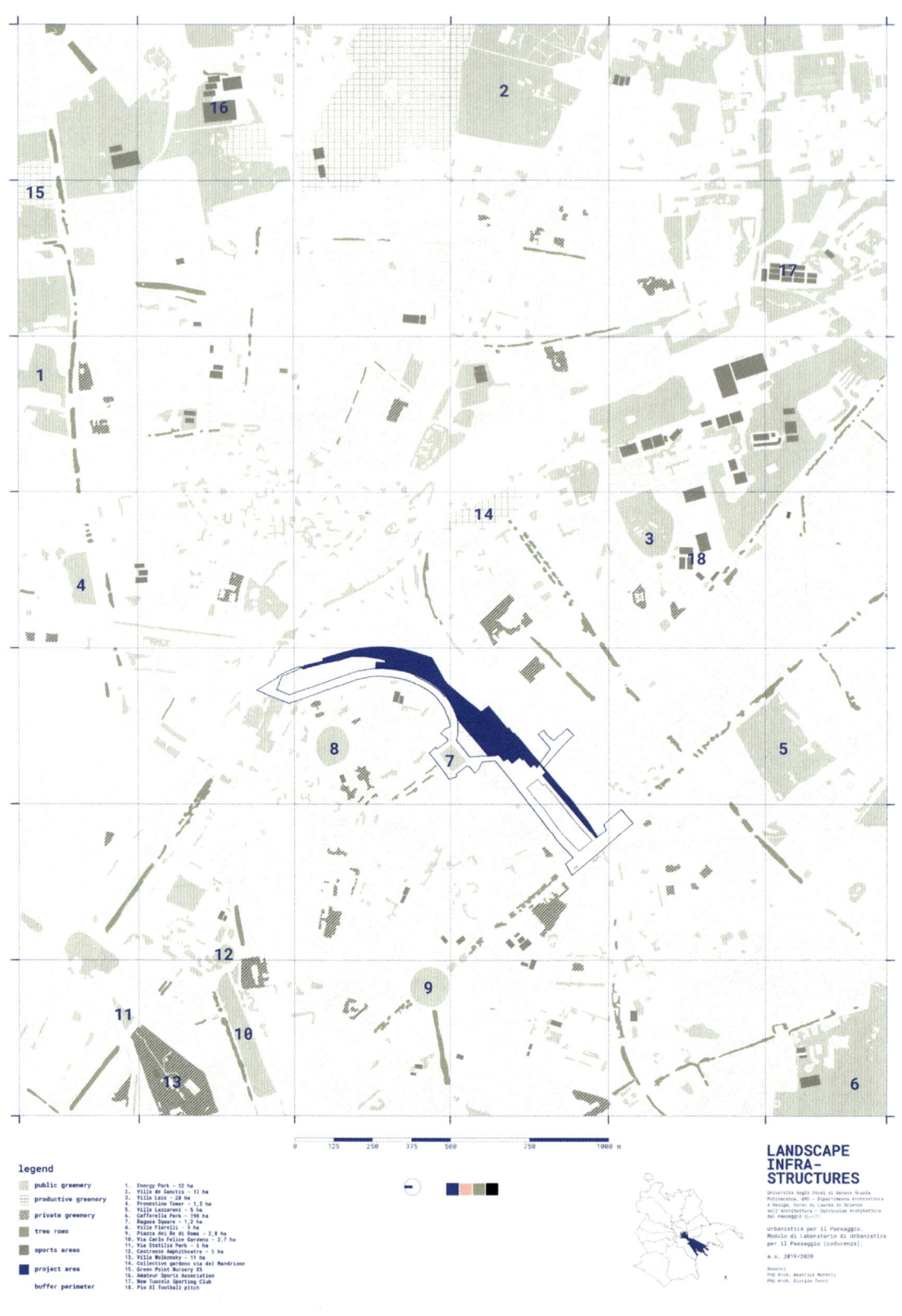

The compact and densified urban fabric around the station does not leave many opportunities for the development of green spaces and in general the area is characterised by a low presence of urban green spaces and parks per capita. The only ones in the vicinity of the station are Villa Fiorelli park, a 9,000 mq public park restored in 2003, and Villa Lazzaroni park. Within the Tuscolano district, worthy of note is the park of Villa Lais, an historic green lung of about 28 ha, an example of a suburban residence dating back to the early 1900s. Here the botanical garden has been adorned with cyclamens, ornamental plants, lavender, roses, oaks and holm oaks, and a small waterfall has been created on a hillock. The area, which used to be a vegetable garden, was instead involved in a redevelopment project that finally transformed it in an equipped park, with additioning an "amphitheatre" aimed to host outdoor events.

VI | GREEN

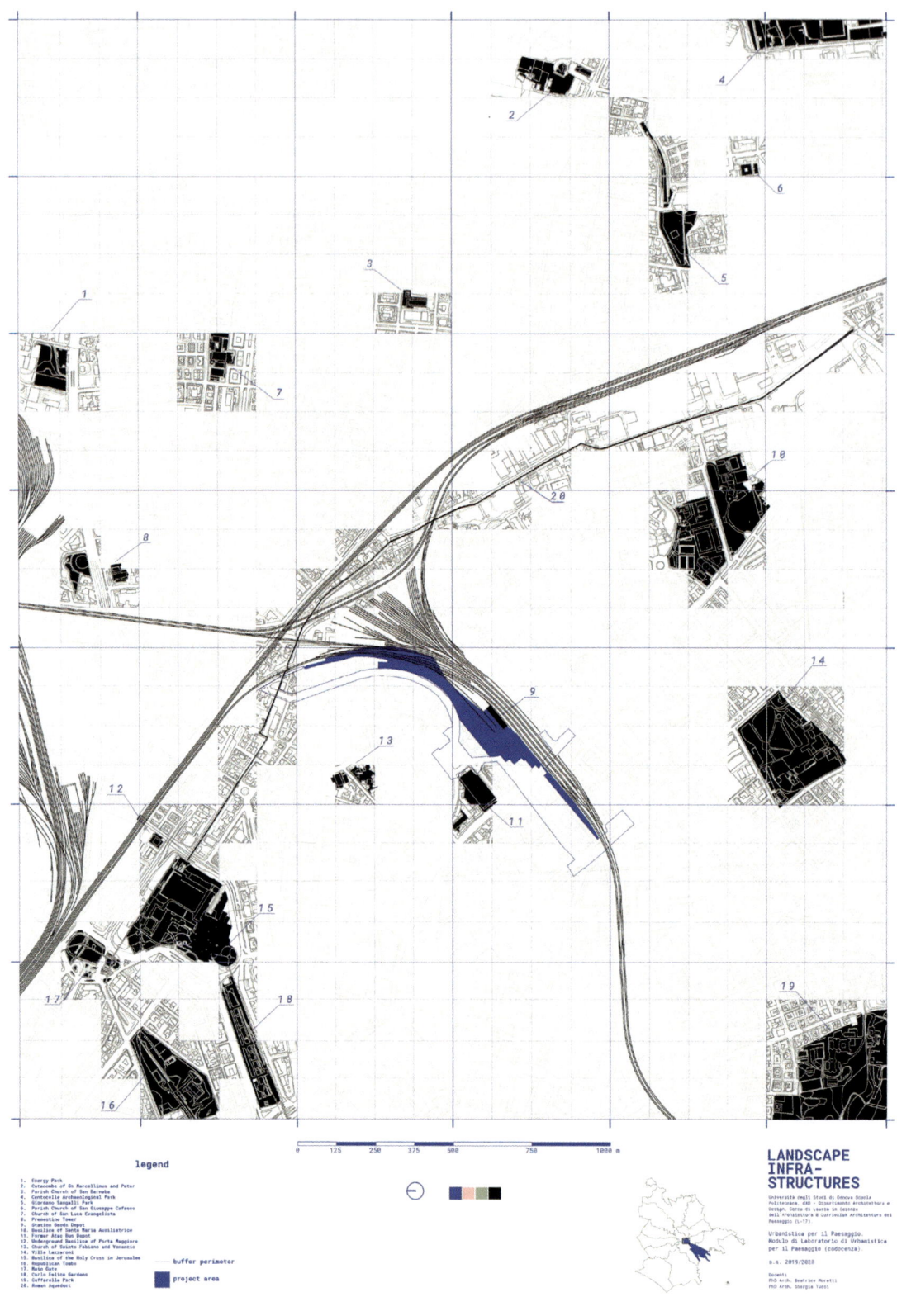

During the years of ancient Rome and the roman empire, the area on which the Tuscolano district stands was an agricultural area, along the roads there were and still are sepulchres and catacombs. During the construction of Piazza Re di Roma, the remains of a Roman villa, which has now disappeared, came to light. Within Via Pescara, two columbaria are still visible today, and near Villa Fiorelli are located the catacombs of San Castulo. The most obvious examples of funerary construction within the area are Via Latina tombs located in a public park and Monte del Grano mausoleum, the largest in the suburb. Another characteristic heritage of the area is the presence of Marco and Claudio aqueducts, later restored by Sixtus V and renamed Felice aqueduct. Nowdays, the ancient Roman Aqueduct "Aqua Claudia" named after Emperor Claudius, not only offers a significant view of the landscape, but also relates to the daily life of the district, intertwining with the railway lines and cultivated fields around which it forms a border.

## VII | HERITAGE

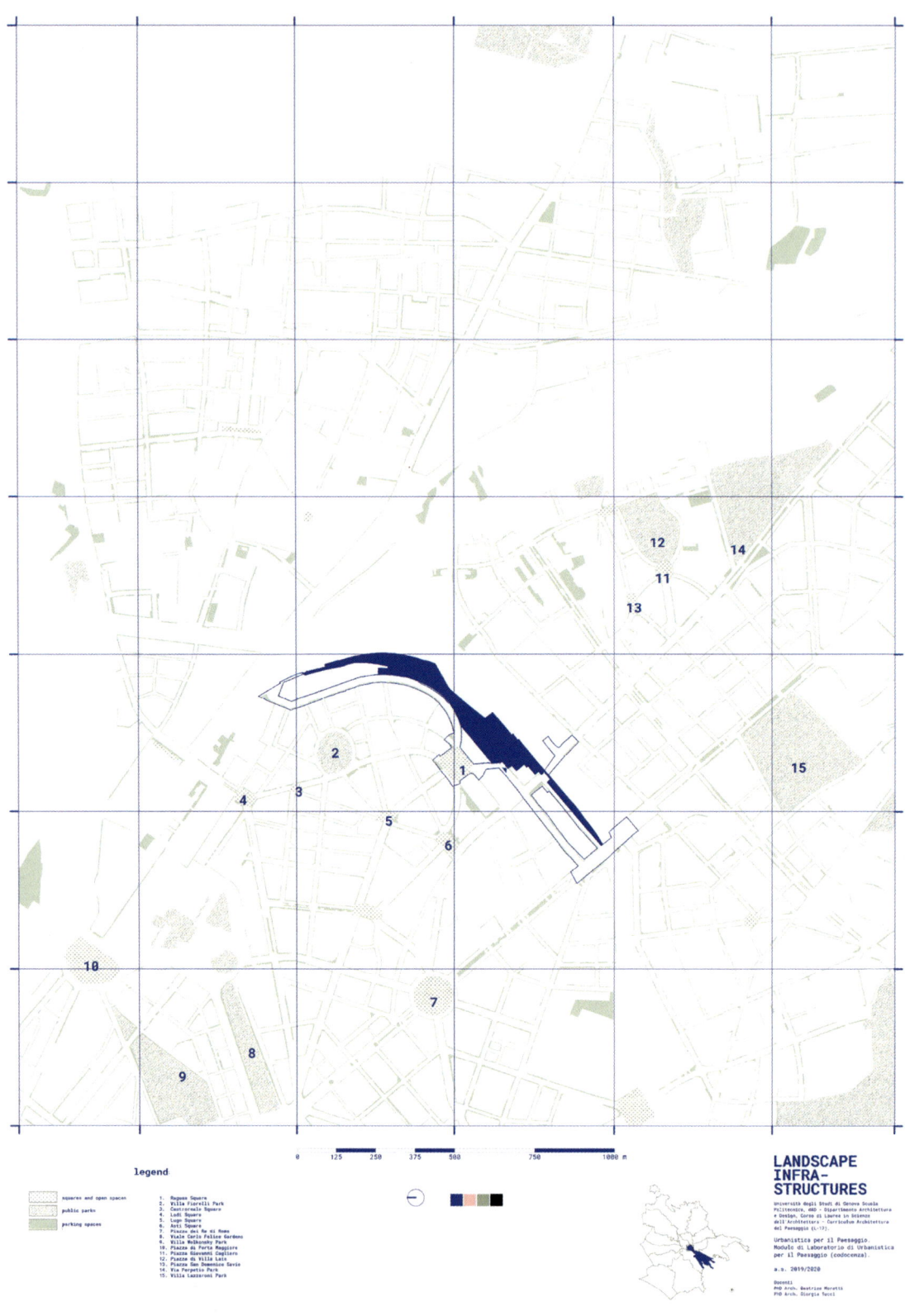

One of the main problems of Tuscolano area is the lack of adequate public spaces and meeting places in line with such a densely populated neighbourhood. The map shows the squares, parks and car parks surrounding the station area, which are relatively undersized compared to the large size of the urban area.

In the Middle Ages the area between today's Piazza Tuscolo, Piazza Re di Roma and Via Taranto was called Trepiccione, this name apparently derived from three sacred images placed along the main road (Tre pizzi). In the 90s, many public spaces were redeveloped (Piazza Lodi, Piazza Ragusa, Largo Michele Unia, Piazza Santa Maria Ausiliatrice), but nowdays much of the area, especially nearby the station, lies in a neglected state of semi-abandonment. For this reason, the Municipality has launched a redevelopment process that will involve Tuscolana Station aimed to restore the unused structures and create new public spaces and ecosystem services (green infrastructure, meeting spaces, temporary shops, fab-lab, sustainable mobility services, etc.).

# VIII | PUBLIC SPACES

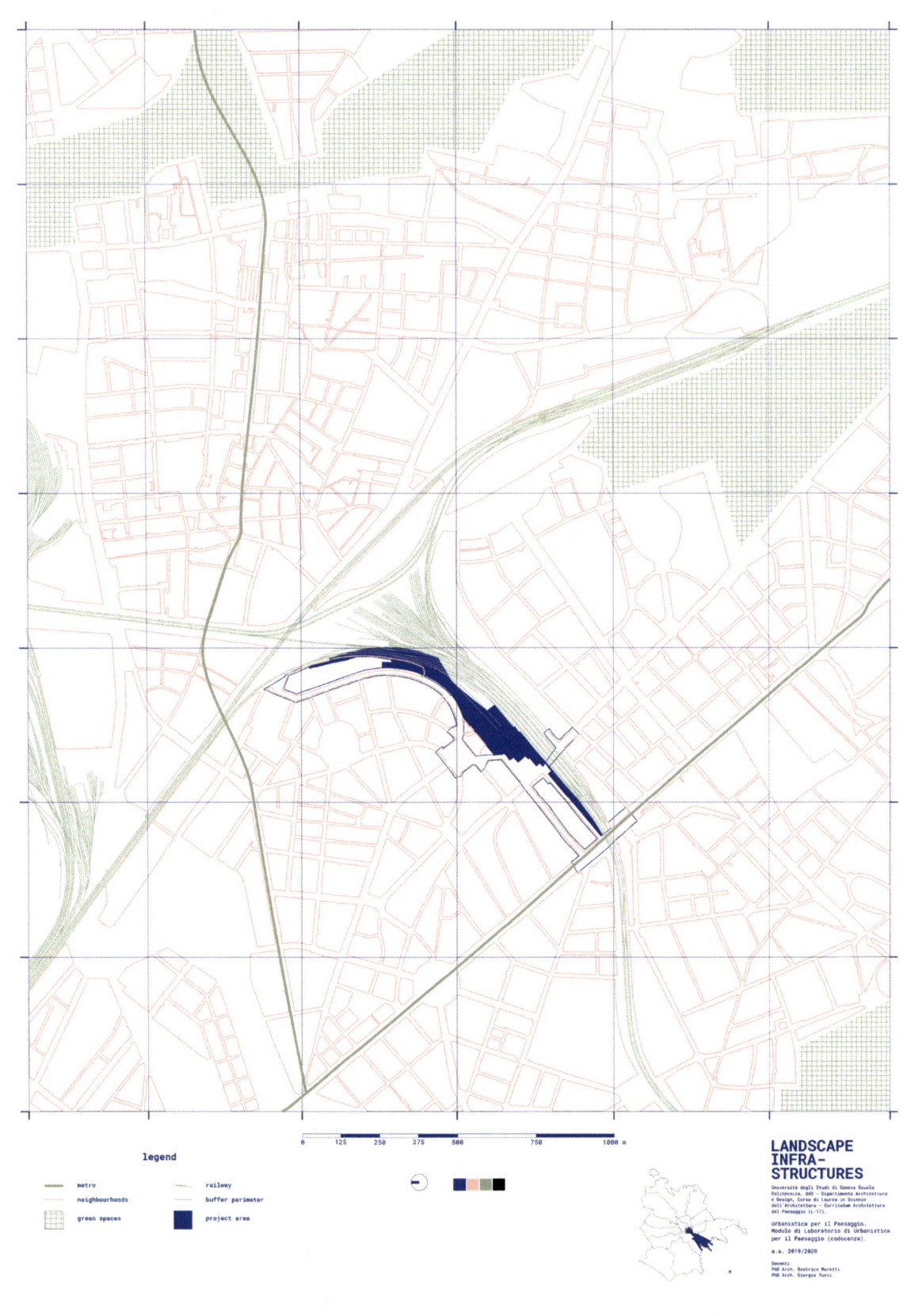

# IX | BORDERS AND INTERFACES

Tuscolana Station area and its perimeter occupies a space of approximately 49,800 m2. The edges of the railway infrastructure are bordered surrounded by: (1) Via Adria, which is parallel to the railway site and is used as a level car park with a housing building, originally intended for railway functions, now unused; (2) the area along Via della Stazione Tuscolana, between the railway site and the urban fabric, consisting of warehouses/depots, some of which are used for storage of railway equipment and for logistics functions; (3) the forecourt facing the station, which is also dedicated to parking.

Access to the station forecourt is through Via Monselice, a side street of Via Tuscolana and Via Mestre. The area of intervention is therefore made up not only of the areas and buildings of the railway station, but also of the station's interchange car park and the small commercial and sports activities adjacent to it. Two clusters of disused or abandoned buildings are concentrated on the edge of the railway line, which are among the objectives of the urban variant "redevelopment and reorganisation of the urban margins along the railway, through the urban development of disused railway areas".

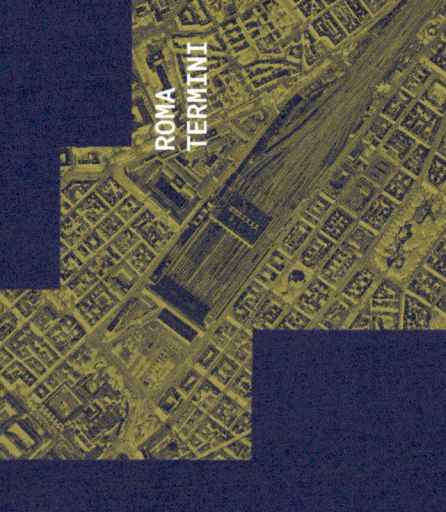

# LANDSCAPE INFRASTRUCTURE [AS]

## *GRAFT*
## *TRANSIT*
## *CROSSING*
## *BRIDGE*

Although secondary to the Capital's intermodal hubs (Roma Termini and Roma Tiburtina), Roma Tuscolana is a large railway node that creates a clear break in the dense Tuscolano district, accessing the urban grid from the south-west. The development of the railway line, which is partly being decommissioned, gives rise to two clear fronts (one to the south-east and the other to the north-west): these are two rigid frames, or urban margins, with which the railway junction establishes relationships of contrast, closure or, instead, potential connection.

For these reasons, the projects collected in the following pages work especially on the theme of accessibility by developing opportunities for transit at different levels: ramps, bridges, suspended walkways that function as grafts on urban and landscape scales. In some cases, these solutions are punctual and solve specific and particularly congested junctions with a single architectural gesture. In other cases, they are solutions linked to height alterations of the ground, in a more pervasive of zero-level intervention approach, especially for pedestrians and cyclists.

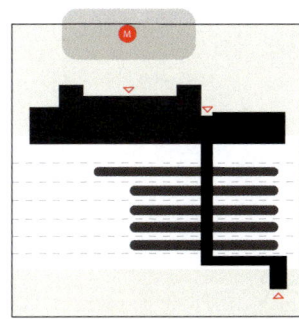

**OSTIENSE**
*transit station*
*overhead connection*

# ROMA TUSCOLANA: NEW ACCESSIBILITY

Starting from an urban analysis of the area adjacent to the Roma Tuscolana station, the project focuses on the theme of accessibility, identifying the main access areas to the area (bridges, walkways, underpasses) where the railroad tracks necessarily constitute a difficult border to cross. Comparing the other stations in Rome (Tiburtina, Trastevere, Termini, Ostiense) with Tuscolana, the different characteristics (neighboring services) and their different behavior towards the issue of accessibility (access points and main connections) are highlighted.

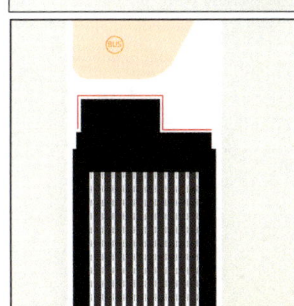

**TERMINI**
*head station*
*direct connection*

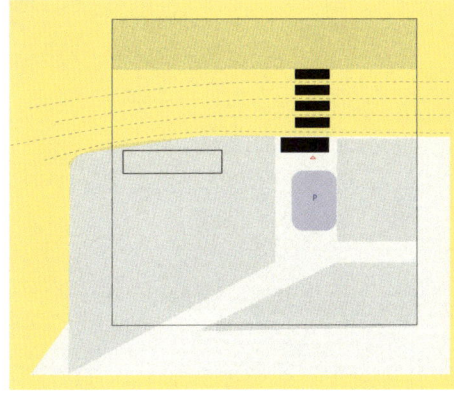

**TUSCOLANA**
*transit station*
*underground connection*

**TIBURTINA**
*bridge station*
*overhead connection*

**TRASTEVERE**
*transit station*
*underground connection*

Rome railway stations.
Author: Andrea Derni.

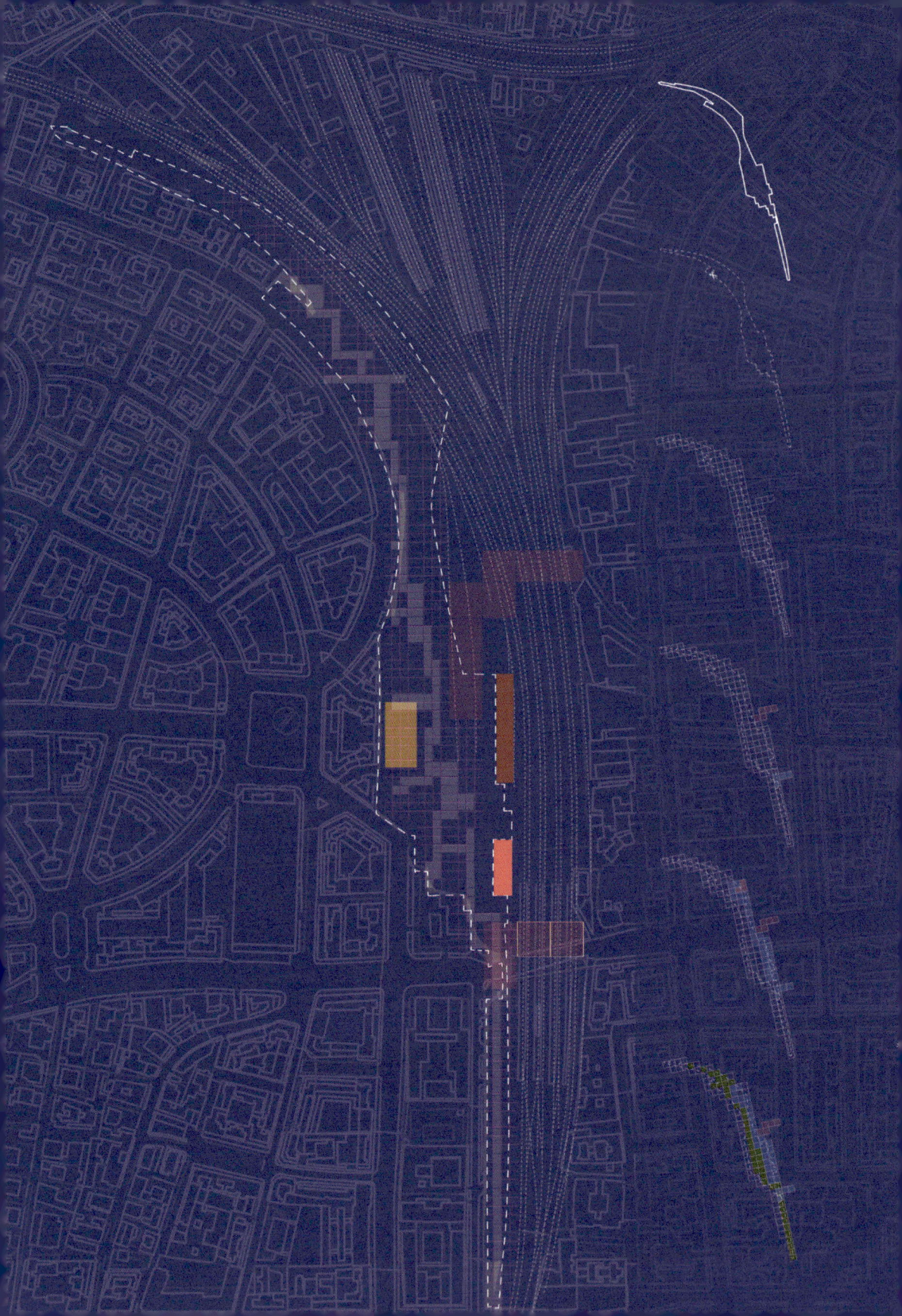

Through the schematic summary it was possible to classify the different connection points:
- in Tiburtina and Ostiense there is an elevated connection that allows access to the stations from one side of the railway tracks to the other;
- at Termini there is a direct connection as it is a head station;
- in Trastevere and Tuscolana there is an underground connection that allows you to arrive at the railway tracks without going beyond them.
The project concept is conceived as a process of gradual contamination of the area. Starting from the area of interest, a highly delimited site, the surrounding barriers have been demolished. Later the area was circumscribed in an ideal centuriation which allowed to create a background grid useful for the dimensioning of the elements. Reconnecting to the main theme, namely that of accessibility, two points of interest have been identified in which to place two walkways in order to create a new variety of connections. Some pre-existing structures - such as the warehouse and the goods shed - have been maintained and converted into a market and restaurant business respectively.

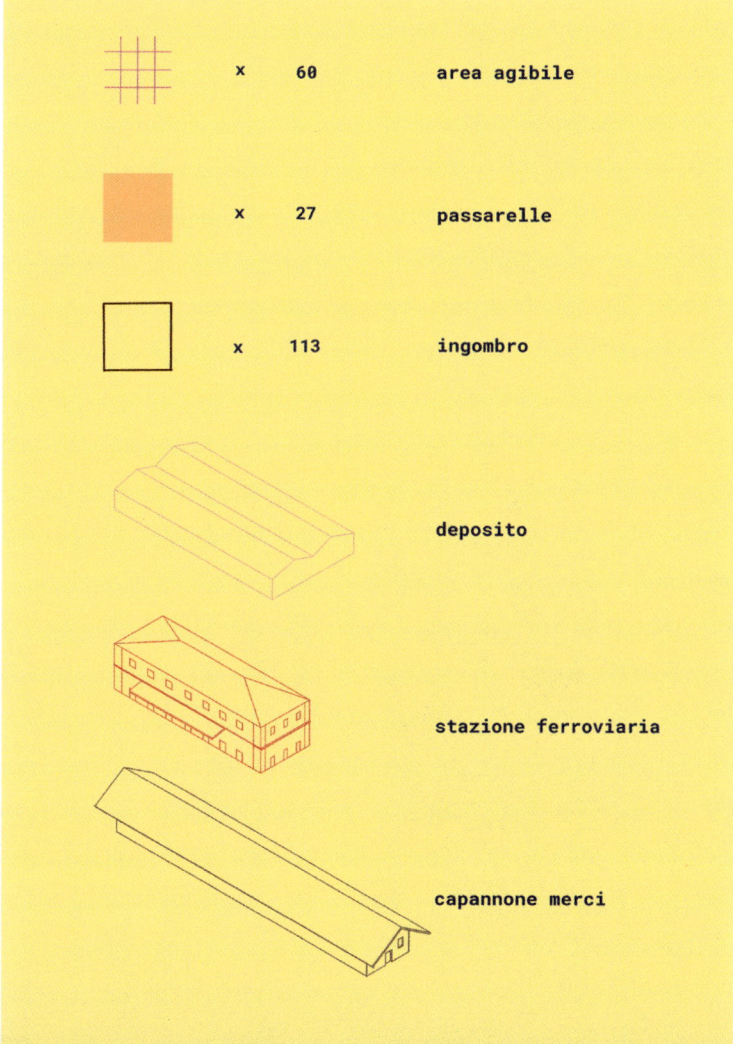

Concept of the project: centuriation and grid.
Author: Andrea Derni.

A.
linear park

In the bottom-up scheme, a cycle / pedestrian path connects Via Adria (metro stop) to the station. In the middle of this connection there is the first ramp which has been conceived and designed on three floors in order to be accessible and usable from the underpass level (Via Tuscolana), from the intermediate level (cycle and pedestrian path) and from the raised level (connection to the platforms). A large square was placed in the central area capable of providing new public spaces to the neighborhood, working on the design of the flooring without filling the space with invasive constructions.

B.
goods shed area
swimming pool

The square thus becomes a place for different mobile and exchangeable scenarios, dedicated not only to average commuters but also to the inhabitants of the neighborhood, overlooked by the warehouse, now the neighborhood market and the goods shed used for catering activities. Some areas of the square are intended for the care and management of greenery, providing for the construction of urban gardens that can be of assistance to the activities present by promoting production at km0.

C.
cycle and pedestrian walk
connection Via Adria - Station

This area is also easily accessible from the other side of the tracks, thanks to an elevated walkway that connects Via Assisi, a road nowadays with no outlet, to the New Piazza or the linear park. The latter connects the New Piazza to Via della Stazione Tuscolana and also consists of zero-level architecture that draws the ground.

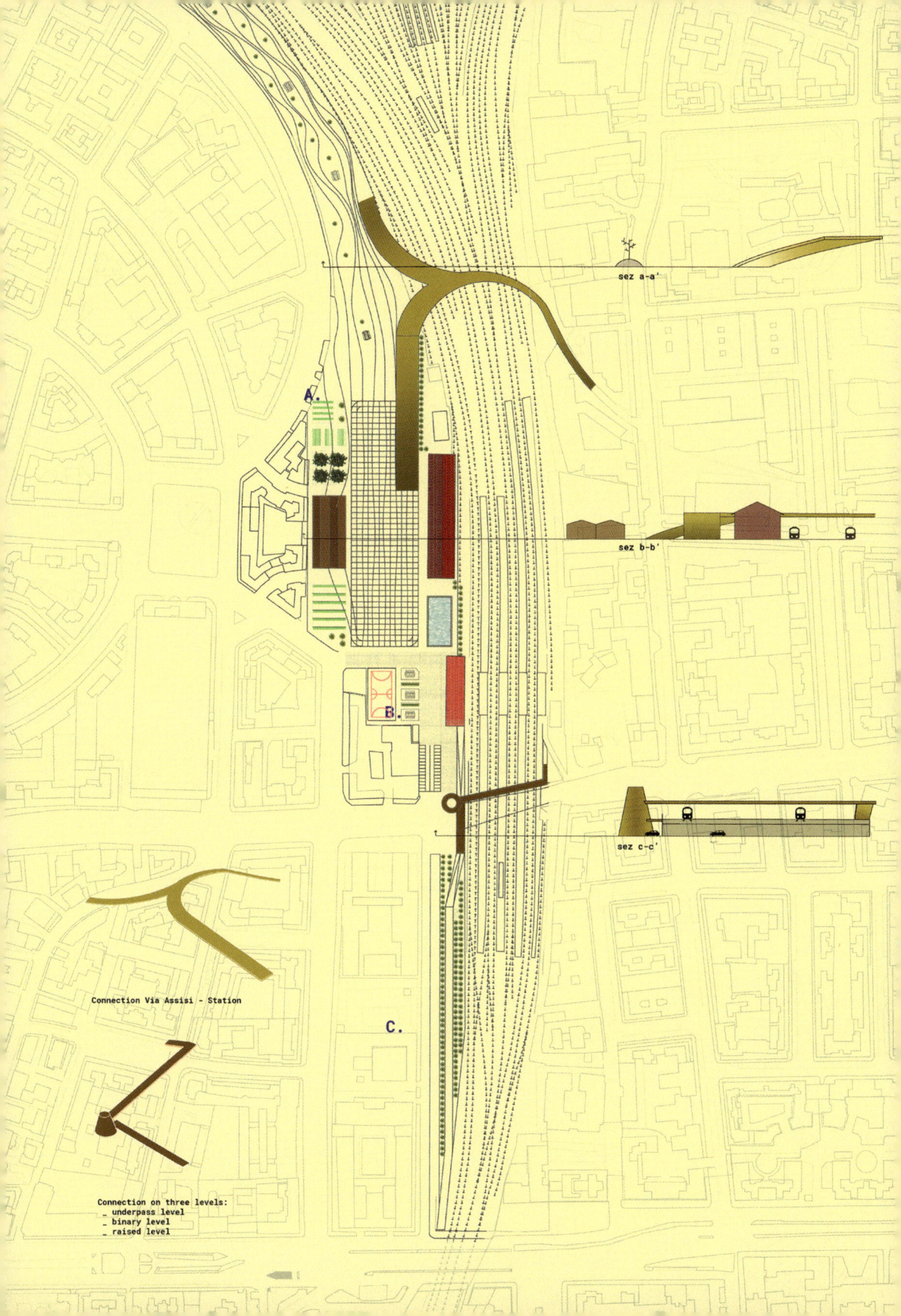

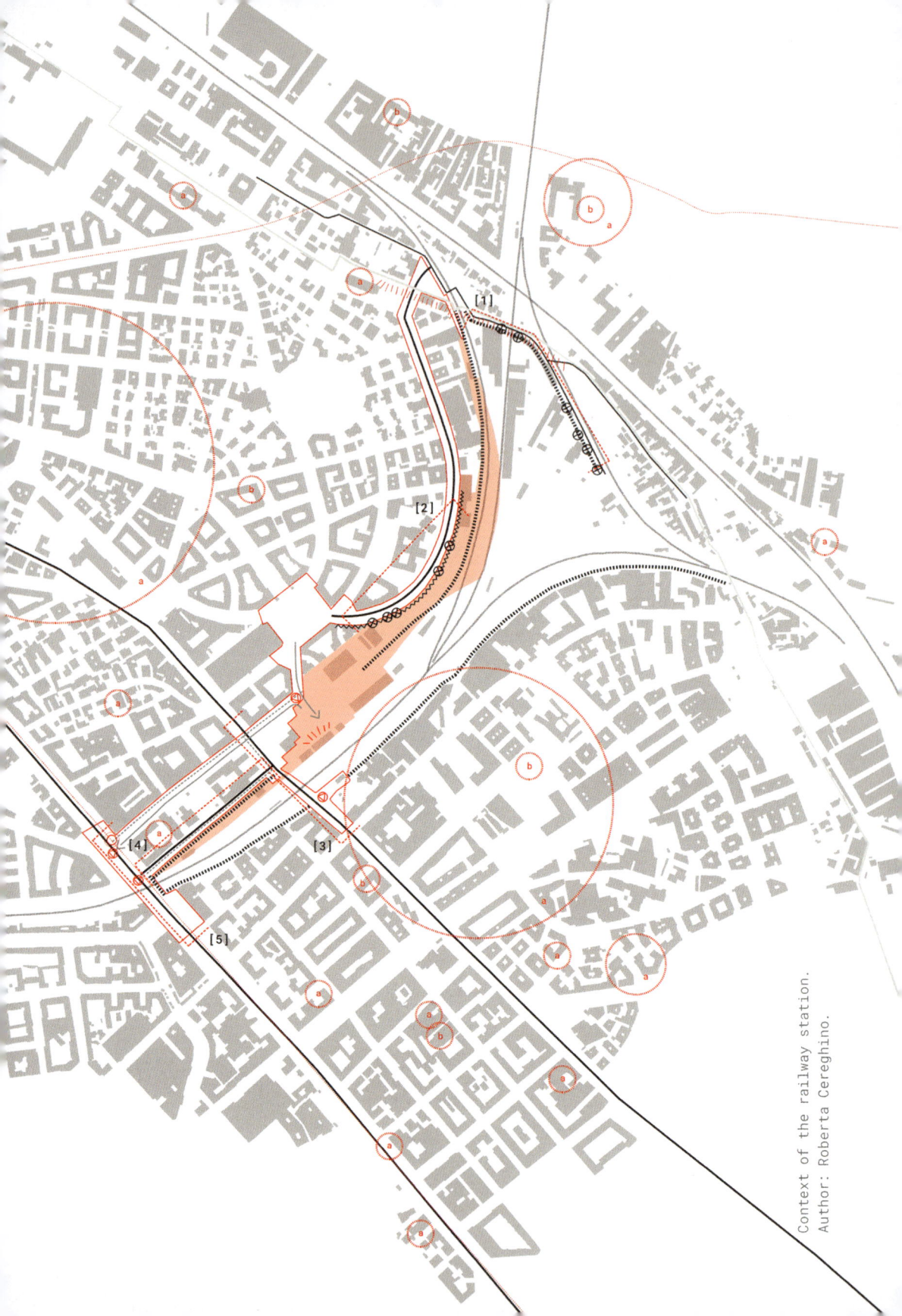

Context of the railway station.
Author: Roberta Cereghino.

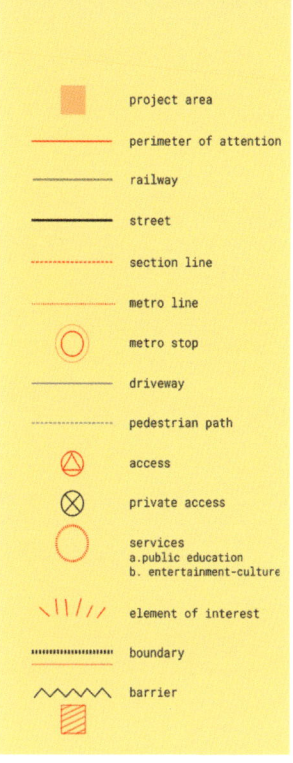

## DIALOGUE BETWEEN THE CITY AND THE RAILWAY YARD

**Railway yard and city ratio:** the railway yard is interpreted as an element of caesura between two edges of the city, an element in its own right (lack of interaction with the surrounding urban fabric).

**Access-barriers-margins:** currently the site has four entrances, one of which is direct to the station, the edges of the yard are identified by the presence of fencing margins, architectural barriers and inaccessible accesses. The area is densely populated with numerous support services.

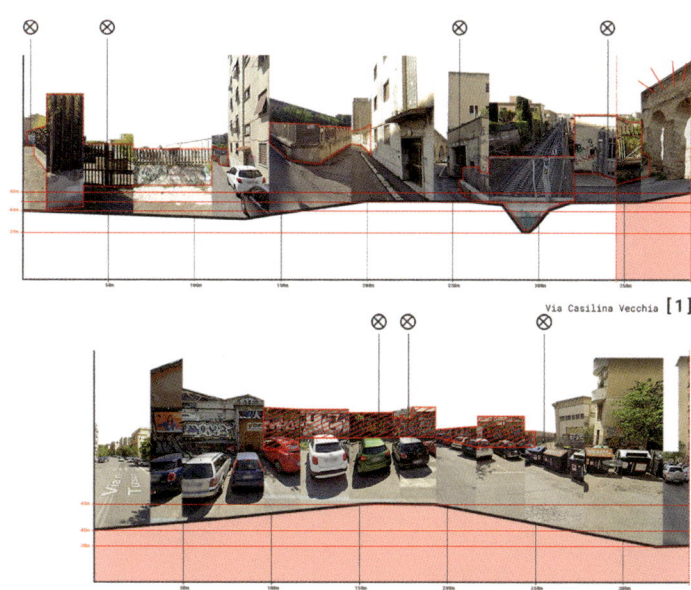

Street level elevation sections around the railway station area.
Author: Roberta Cereghino.

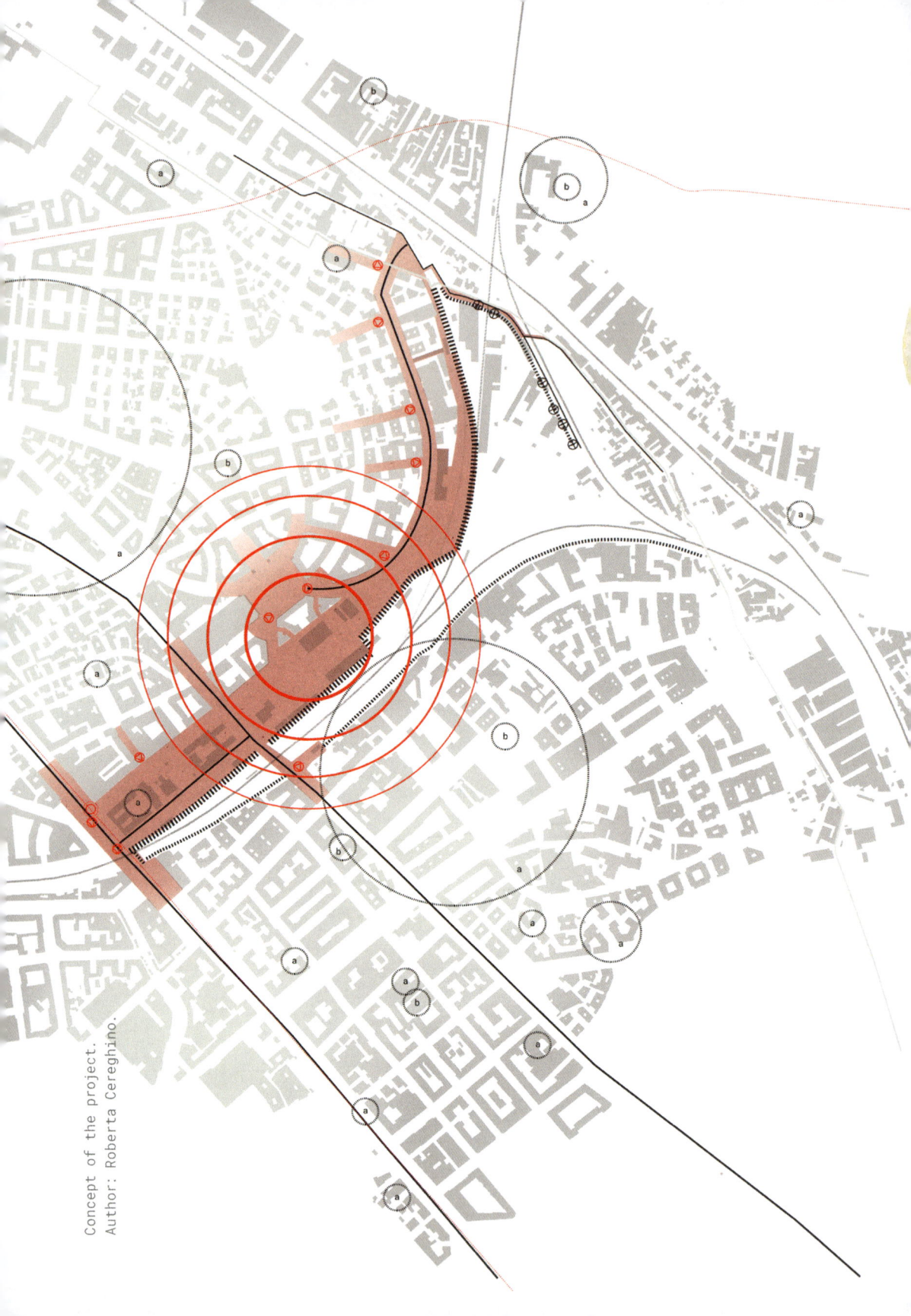

Concept of the project.
Author: Roberta Cereghino.

**Representation of the project area on three levels:**
- the first level is the state of the art and identifies the edge, generated by the railway, with its barriers and margins (the black lines);
- the second represents the scheme of the hypothetical new accesses generated by the "opening" of the edge in correspondence with the existing cross roads to the railway which connect to the urban fabric;
- the third drawing indicates the project area with its new extensions towards the city and the centralization of services within it.

Representation of the project area on three levels.
Author: Roberta Cereghino.

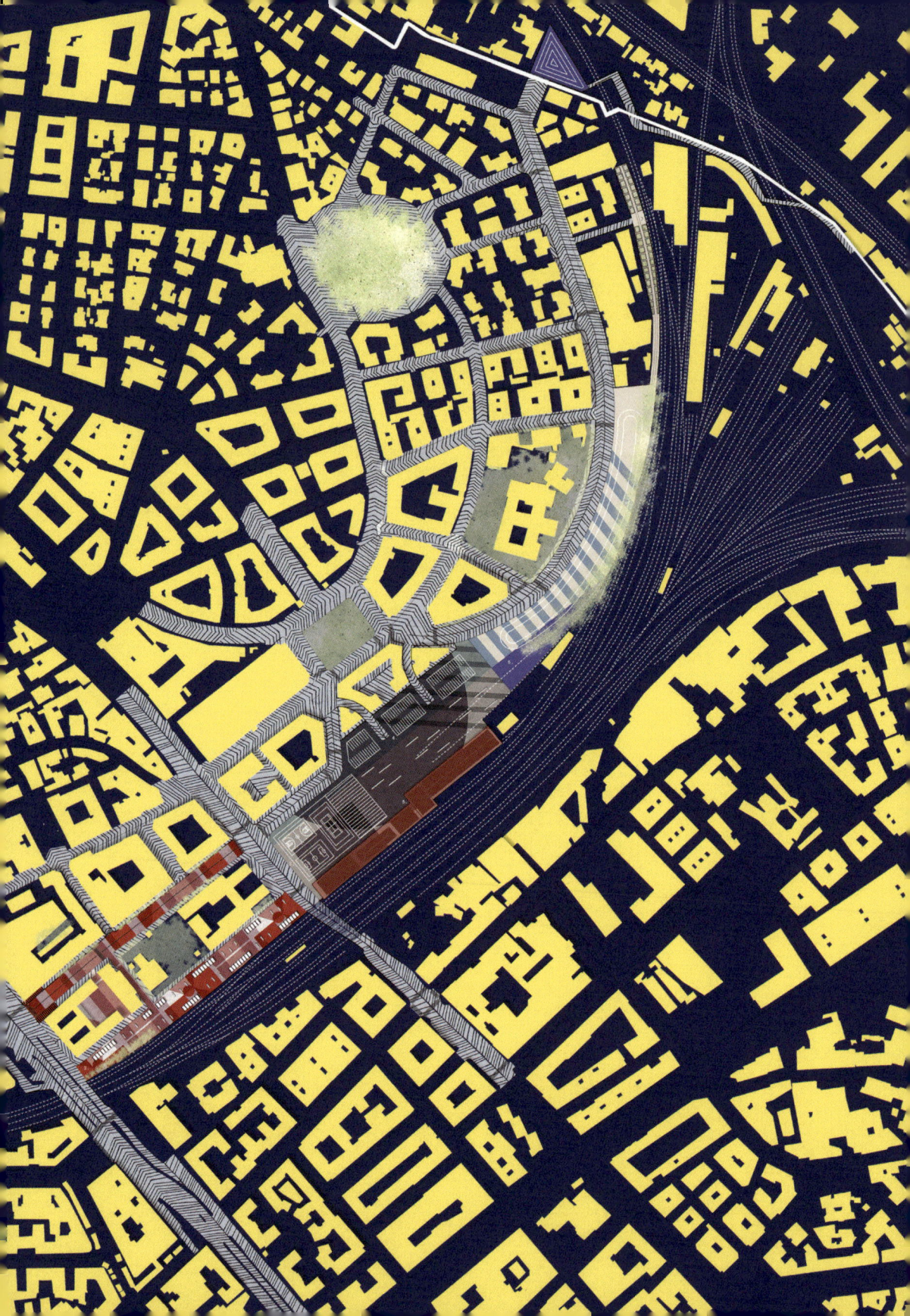

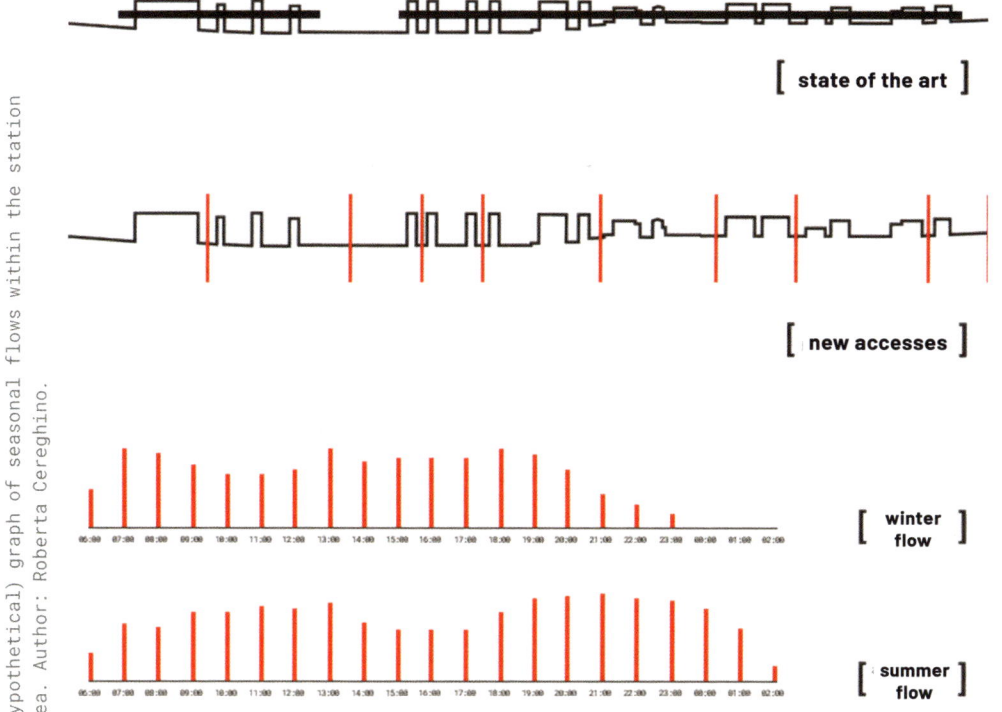

(Hypothetical) graph of seasonal flows within the station area. Author: Roberta Cereghino.

Vision of the railway stations. Author: Roberta Cereghino.

**Vision**
Reconfiguration of the area through the ground design of new spaces which suggest use and accommodate multiple functions simultaneously. In addition to the multifunctional structure that encompasses the station building and the warehouse, the fixed elements are in a minimum quantity in order to make the spaces as free as possible and therefore flexible in accommodating temporary activities.

The project areas are:
- exhibition area;
- plyground;
- event area parking;
- sports activity area;
- urban gardens;
- didactic activity area.

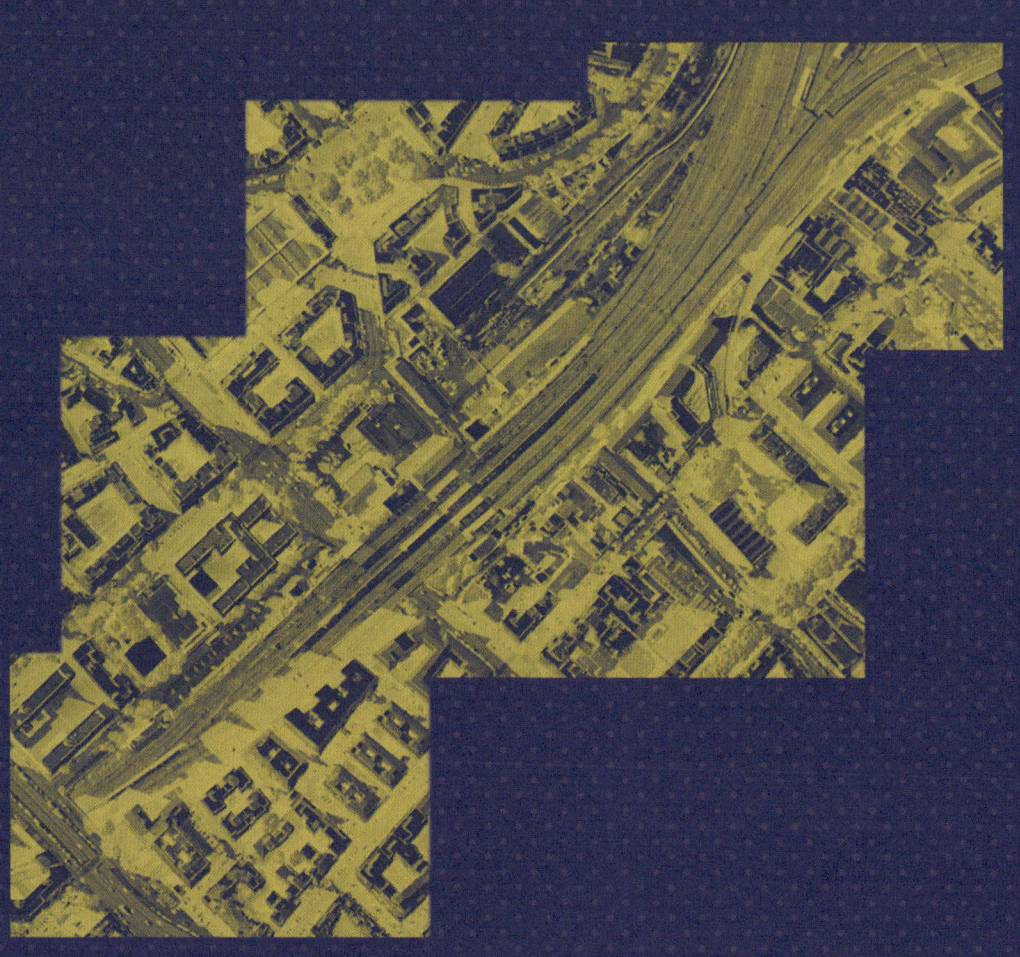

# LANDSCAPE INFRASTRUCTURE [AS]

## *HINGE*
## *PLATFORM*
## *INCUBATOR*
## *GRID*

Roma Tuscolana is a metropolitan railway pole. It is used for urban and suburban transports, for intermodal connections with the metro line and for transfers to and from Fiumicino airport. Although its longitudinal layout interrupts the city grid creating a decisive *caesura*, the contemporary uses and flows of Roma Tuscolana make it an essential urban connection node.

The strategic theme addressed by some projects was therefore the connectivity of the Tuscolana station, on the one hand with the city's public transport infrastructure network, and on the other with the city's historical and cultural heritage network.

The projects collected in the following pages work on the idea of an infrastructural link: a platform, an incubator of services, a dynamic grid capable of transforming the station into a hub of daily aggregation.

In some cases, the projects have developed solutions linked to the system of public and green spaces, proposing the railway station as a green mobility generator. In other cases, the connectivity of the context has been stimulated through the insertion of punctual connection devices (a market, a theatre, social and collective spaces, an arena, etc.) or the reinterpretation of the architecture of the Roman aqueduct as a multifunctional and multidirectional urban and territorial infrastructure.

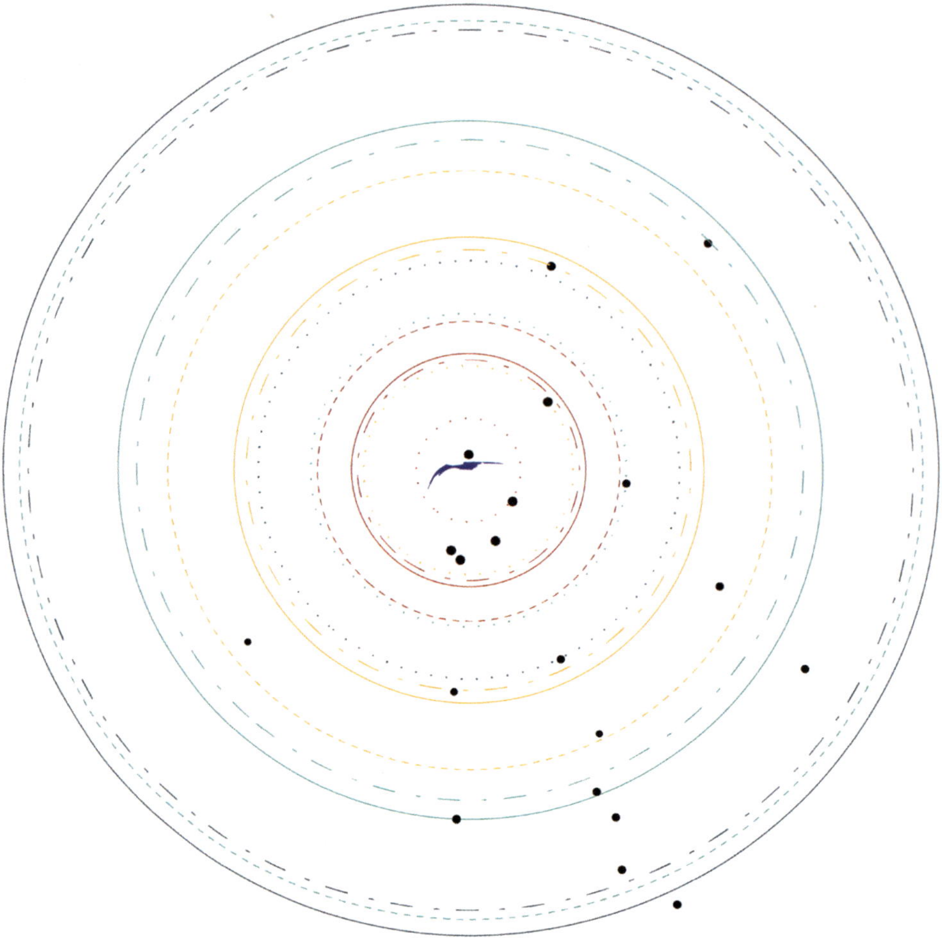

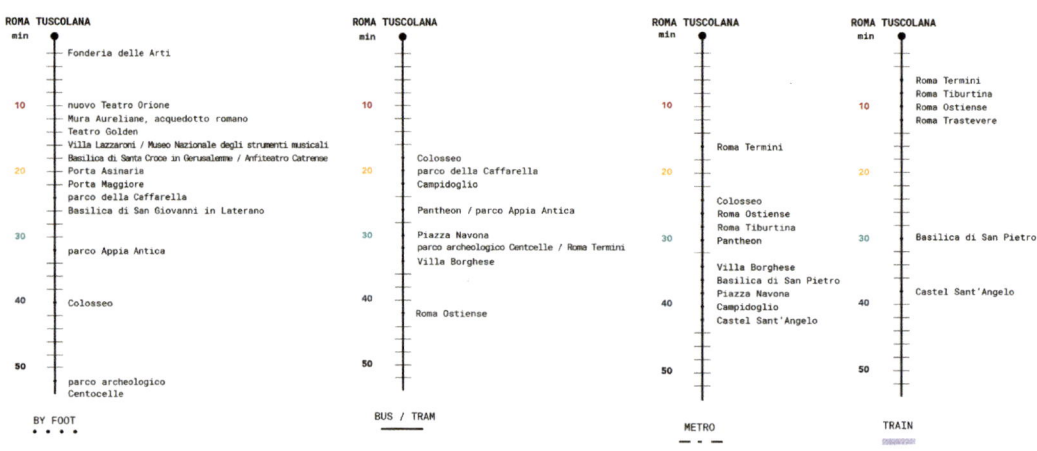

## ROMA TUSCOLANA. A DIFFUSED CULTURAL PARK

The railway station of Rome Tuscolana has the significant characteristic of being located in a part of the city where historical monuments, cultural and natural areas are easily accessible. It is therefore an urban connection node. In this case it has been studied at a temporal level as well as urban planning in relation to the Roman historic center. In the study, the isochrones indicate the travel time required from the Rome Tuscolana railway station to reach the areas of interest selected according to the public transport used. The infrastructural network of city public transport thus becomes the key to urban connection between the project and the surrounding areas.

Left: Timeline diagram showing travel times from Rome Tuscolana station and historical and architectural emergencies
Right: Abacus of historical and architectural emergencies.
Authors: Andrea Oriati, Elisa Pizzorno, Arianna Viola.

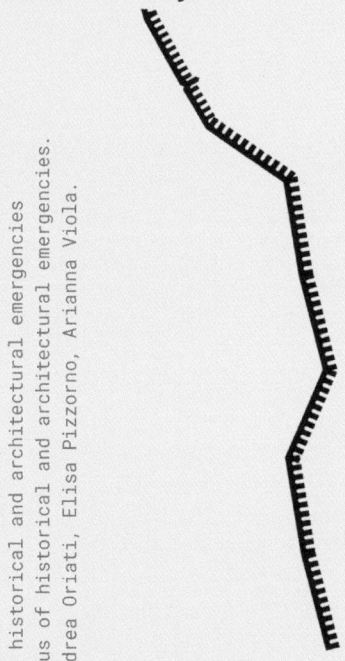

Teatro Golden / Teatro Orione

Museo Nazionale degli strumenti musicali

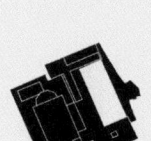
Anfiteatro Catrense / Basilica Santa Croce

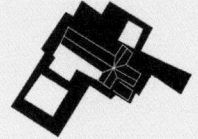
Basilica di San Giovanni in Laternao

Fonderia delle Arti

Villa Lazzaroni

Campidoglio

Colosseo

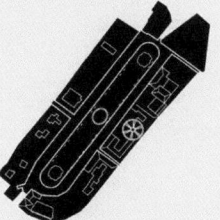
Piazza Navona

Pantheon

Mura Aureliane acquedotto romano

Castel Sant'Angelo

Basilica di San Pietro

1_ parco archeologico Centcelle
2_ parco Appia Antica
3_ parco della Caffarella
4_ Anfiteatro Catrense
5_ Villa Lazzaroni
6_ Mura Aureliane, acquedotto romano
7_ Fonderia delle Arti
8_ nuovo Teatro Orione
9_ Teatro Golden
10_ Museo Nazionale degli strumenti musicali
11_ Basilica di Santa Croce in Gerusalemme
12_ Porta Asinaria
13_ Basilica di San Giovanni in Laterano
14_ Colosseo
15_ Campidoglio
16_ Pantheon
17_ Villa Borghese
18_ Piazza Navona
19_ Castel Sant'Angelo
20_ Basilica di San Pietro
21_ Roma Tiburtina
22_ Roma Termini
23_ Roma Ostiense
24_ Roma Trastevere
25_ Porta Maggiore

The identification of the main urban road axes that interrupt or cross the project area has resulted in an orthogonal spatial grid on which the design concept is organized.

The routes and the punctual devices, as tools of connection to the city and of direct connection with the transport system, are inserted respectively on lines and nodes of this grid.

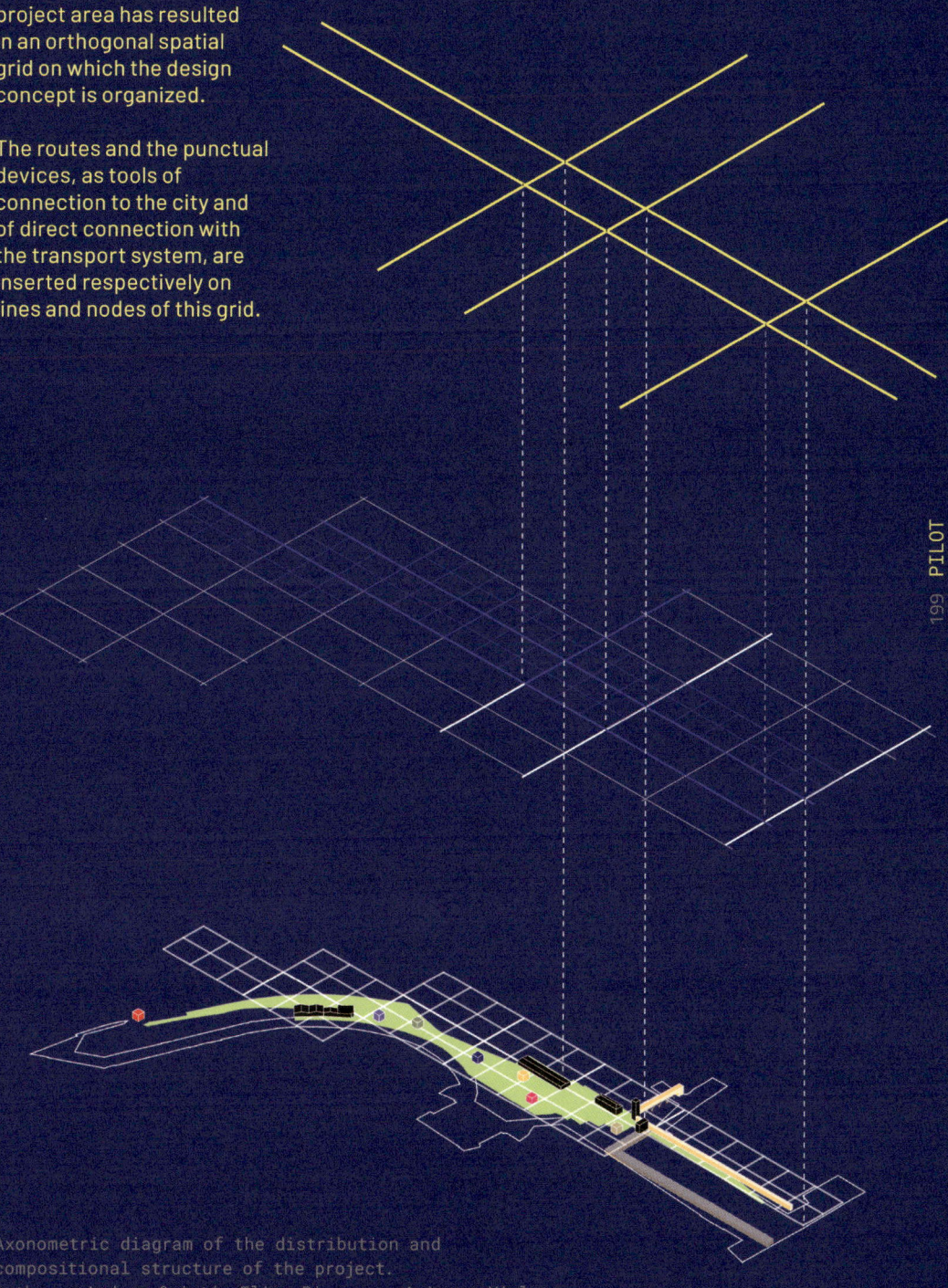

Axonometric diagram of the distribution and compositional structure of the project.
Authors: Andrea Oriati, Elisa Pizzorno, Arianna Viola.

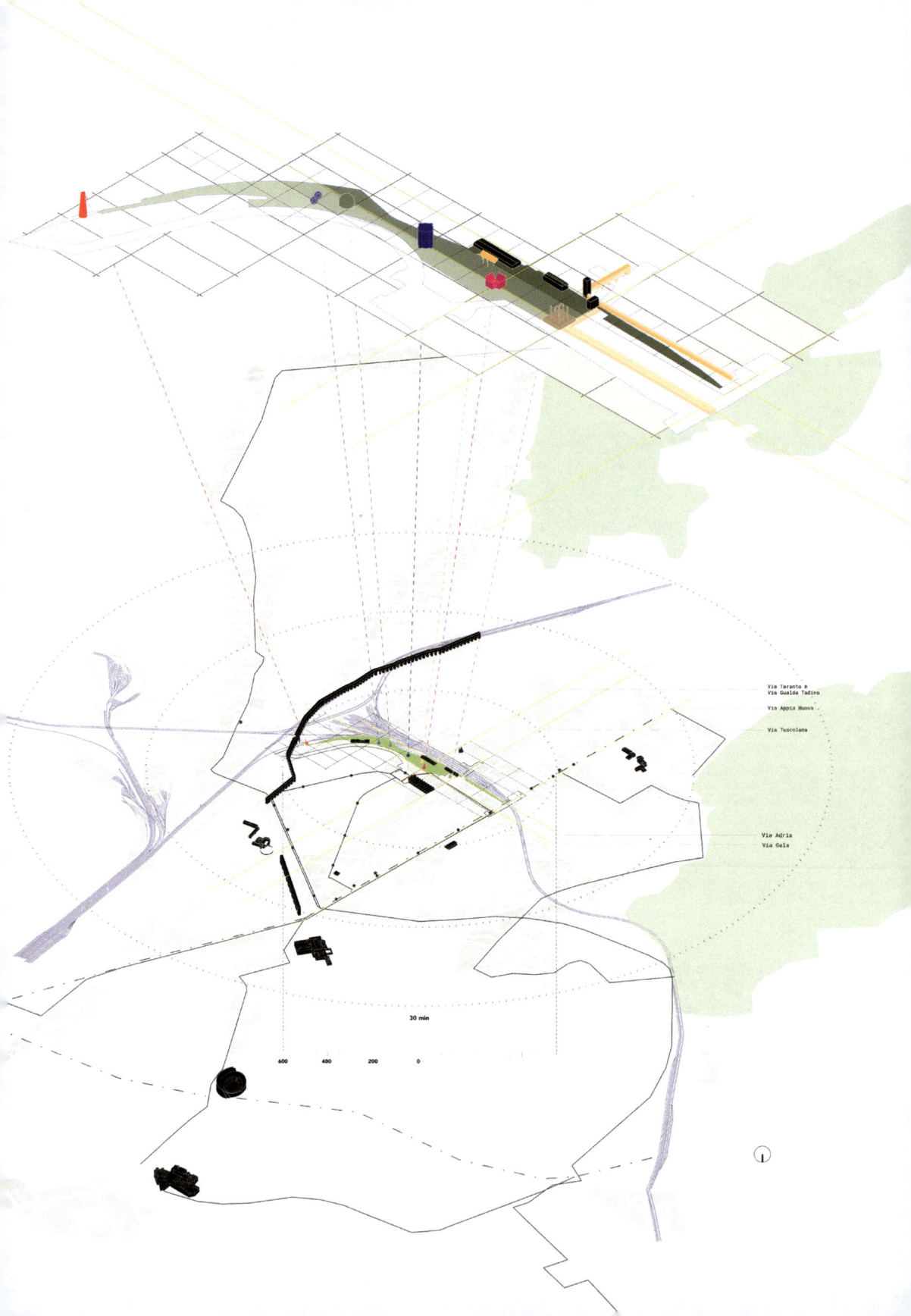

Reworking of the functional and symbolic role of emergencies applied in the design of devices.
Authors: Andrea Oriati, Elisa Pizzorno, Arianna Viola.

The urban hub of Rome Tuscolana today represents an axis of separation between two parts of Rome which, although distinct, contain elements of environmental, historical, artistic value , which would offer the opportunity for them to dialogue with each other.

The analysis of the area is based on the determination of the travel times it takes from the Roma Tuscolana station to reach these elements up to the city center. The emergencies enclosed in the thirty-minute walk isochronous, in particular, have been the object of interest for the development of the project, as they are disconnected and undervalued despite their position close to the station, a place of passage, dynamism and community.

The identification of the main urban road axes that interpose or cross the project area has resulted in an orthogonal spatial grid on which the design concept is organized. The routes and the punctual devices, as tools of connection to the city and of direct connection with the transport system, are inserted respectively on lines and nodes of this grid.

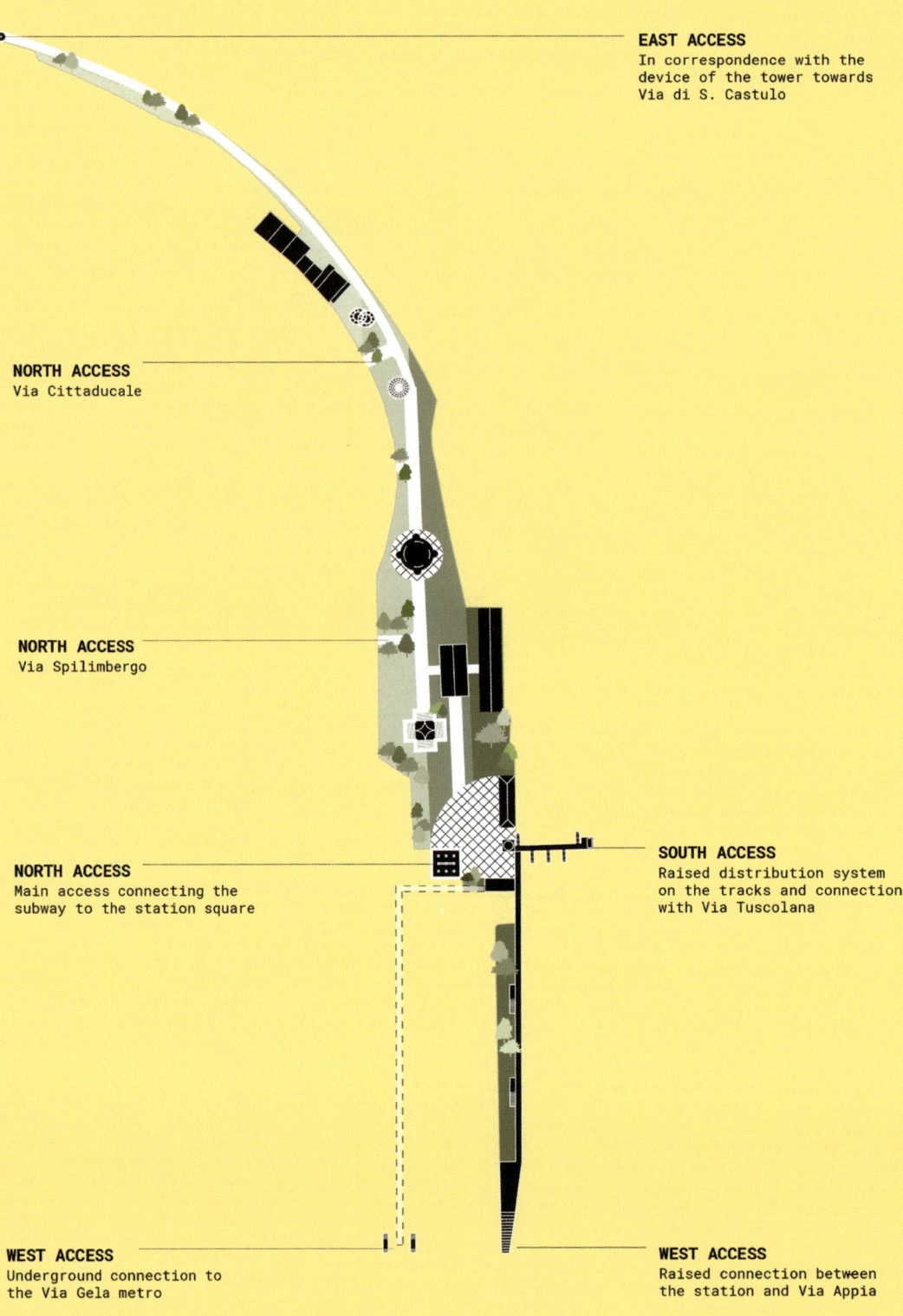

The devices included in the project area represent in an iconic way the main features of the area at a functional but also monumental level. They outline the main entrances and paths, functionally connect the area to the cultural, artistic, sporting and scientific activities present both in the Tuscolano district and in the rest of the historic center. The Diffuse Cultural Park of Rome Tuscolana is therefore the opportunity to include the neighborhood in the historical and cultural network of the city and the railway station of Rome Tuscolana the infrastructural link.

Abacus of devices.
Authors: Andrea Oriati, Elisa Pizzorno, Arianna Viola.

SWOT analysis of the project area.
Authors: Cecilia De Mattia, Cassi Randivoson.

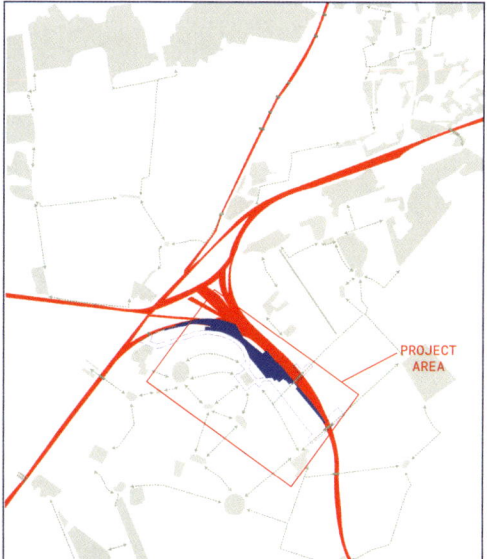

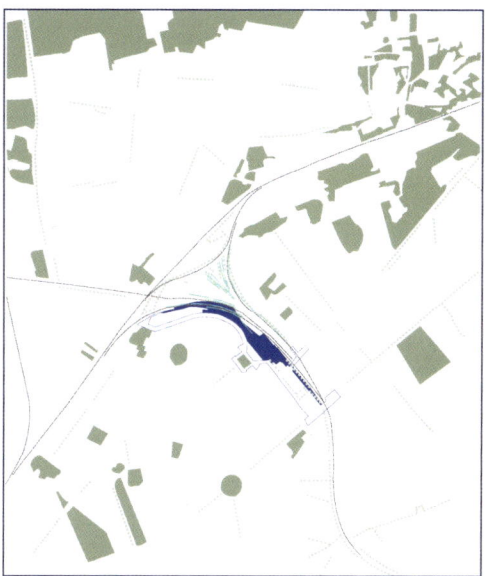

1. Piazza Ragusa
2. Piazza/Parco di Villa Fiorelli
3. Piazza Castroreale
4. Piazza Lodi
5. Piazza Lugo
6. Piazza Asti
7. Piazza dei Re di Roma
8. Giardini di Viale Carlo Felice
9. Parco di Villa Wolkonsky
10. Piazza di Porta Maggiore
11. Piazza Giovanni Cagliero
12. Parco di Villa Lais
13. Piazza S. Domenico Savio
14. Parco Via Perpetio
15. Parco di Villa Lazzaroni

## STATION AS A GREEN MOBILITY AGGREGATOR

Through an in-depth study of the typical traveler profile of the Rome Tuscolana railway station, it was analyzed that the station is not seen as a meeting place but only as an area of arrivals and departures. Although there is an impractical stretch of cycle path, the station is more reached through active mobility (bicycle, scooter, skateboard, etc.) than by public or private transport. The project idea is therefore to implement a redevelopment of the area that focuses precisely on increasing this type of mobility, through the creation of multifunctional routes that follow the course of the track layout. By re-appropriating the disused tracks in the area, a new gymnastic path was created, while the other paths, differentiated by a change of flooring, were used as a bike, skateboard and skating area. The station building has been readapted as an aggregation center through the inclusion of a rest area with bicycle parking spaces with integrated vertical green structures.

### STRENGTH
- facilitated transversal crossing with underpasses
- presence of numerous pedestrian paths and various urban green areas
- predominant active mobility

### WEAKNESSES
- lack of cycle path and bike parking
- presence of various green areas and fragmented paths
- lack of meeting space
- lack of direct passages between the north-south area and the metro-station connection
- railway as an architectural barrier

### OPPORTUNITIES
- presence of disused tracks
- station as an aggregative pole
- thickening of the existing green infrastructure

### THREATS
- north-east area without sidewalks
- presence of abandoned buildings

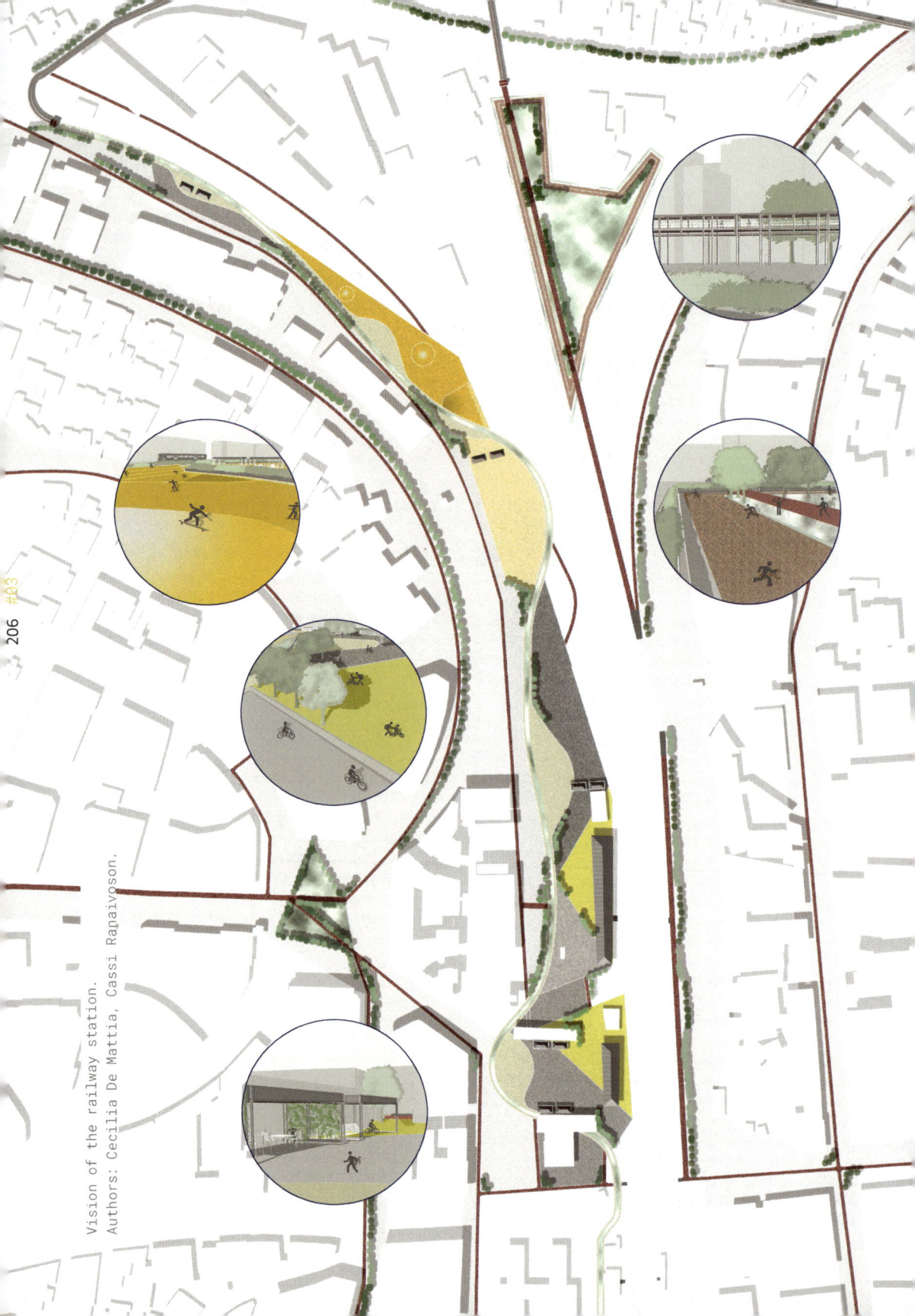

Vision of the railway station.
Authors: Cecilia De Mattia, Cassi Rapaivoson.

**The railway as a barrer:** The Rome Tuscolana railway is seen as a barrier to be demolished by inserting bridges and underpasses.

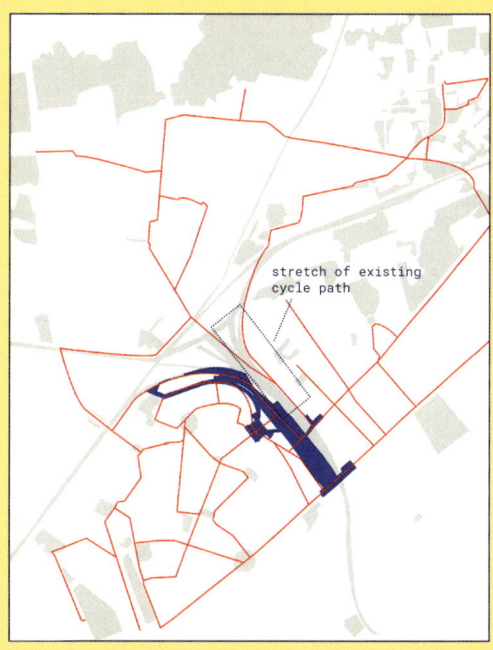

**Extension of the cycle path:** Enhancement of active mobility by extending the current very weak cycle path.

**Station as an aggregative pole:** Redevelopment of the station area from a place of passage to a lived-in place through the inclusion of a new interchange node for active mobility.

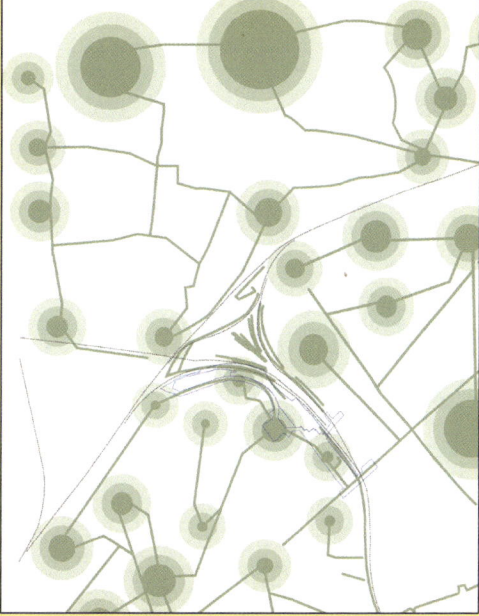

**Strengthening of green infrastructures:** Development of a green path from which various multifunctional spaces are created that reconnect the pre-existing open spaces.

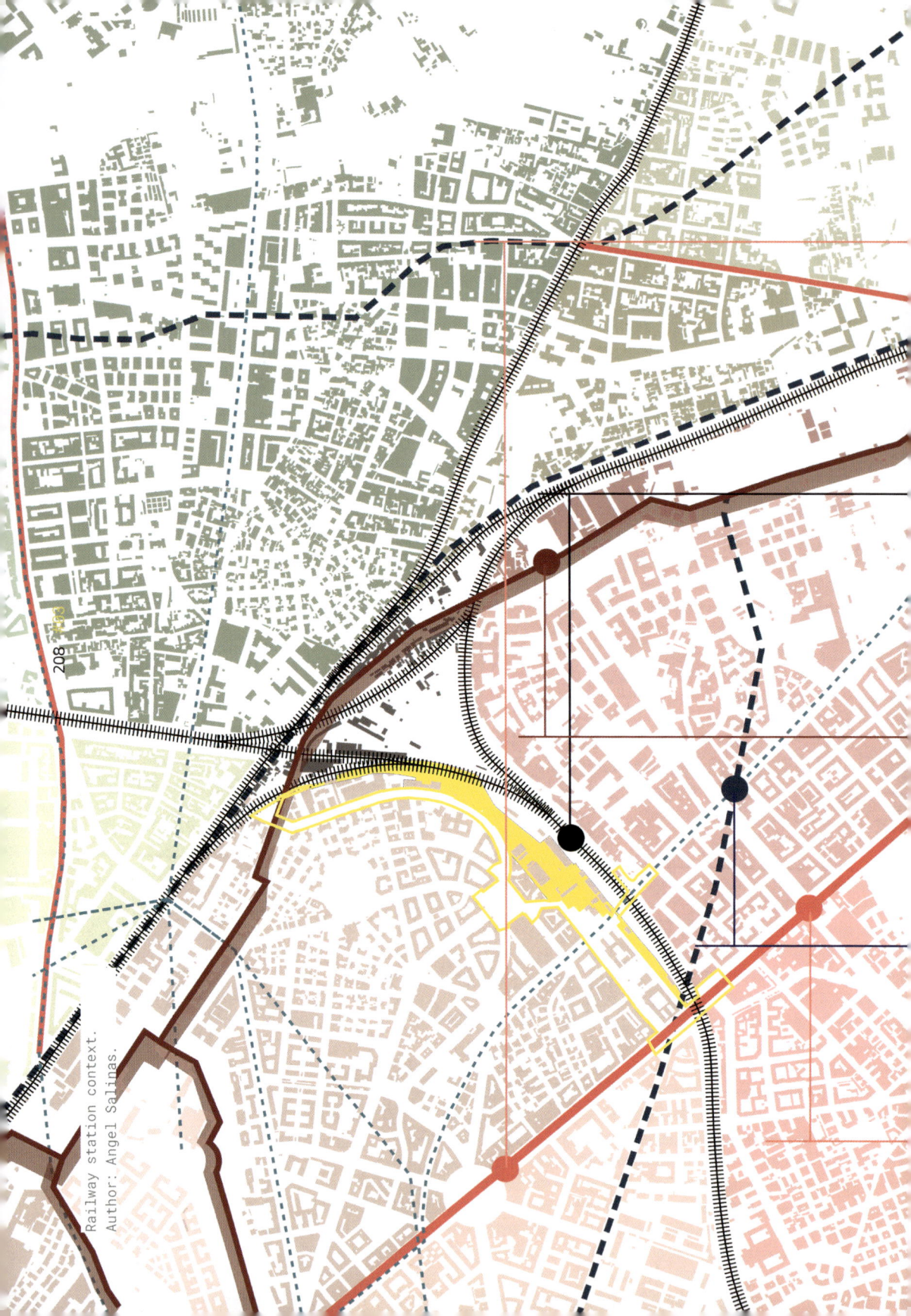
Railway station context.
Author: Angel Salinas.

# CONTEMPORARY AQUEDUCT: FROM THE EDGE TO THE URBAN SUTURE

## Framework

Rome is composed of a myriad of longitudinal axes, which have defined its current urban landscape and configuration. These axes are mobility or transport, historical cultural and hydrographic. In the Tuscolana area there is evidence of a notorious urban fragmentation due to these axes, as they have physically or virtually separated two or more spaces from each other, developing the role of "urban edge". In the analysis tangible axes are identified, which are exposed above level 0 and alter the urban dynamics of the area, such as the railway, the Appian Way and the historical aqueduct. Also, the intangible axes, understood as those axes that do not have direct contact at level zero; these axes are subway, such as the metro line, the hydrographic canals and the subway aqueducts.

## Roman aqueduct

The project is generated from the analysis of the Roman aqueduct, defined as a longitudinal axis, sometimes as an edge and sometimes as a suture; and also as a tangible and intangible element that depends on external physical conditions.
It presents a play of full and empty spaces on both axes. The emptiness of the upper part is identified as a conductive element, which guides and the emptiness of the arches as a connecting element, which allows both sides to have physical and visual contact. By way of synthesis, the aqueduct is defined as a conductive, connecting and historical patrimonial element; which generates and allows the displacement in two axes; the longitudinal and transversal. The proposed urban strategies are developed following guidelines obtained from the analysis of the aqueduct. First, the identification of the longitudinal green areas as a connecting element. Then, it is proposed a network of roads that have as a centrifugal axis the project area.

Road axes

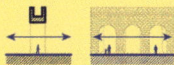

Railway

Aqueduct

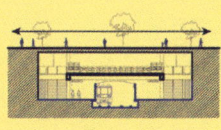

River channel

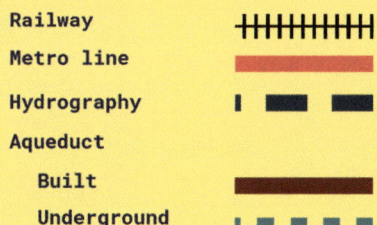

Metro

Concept of the project and new links.
Author: Angel Salinas.

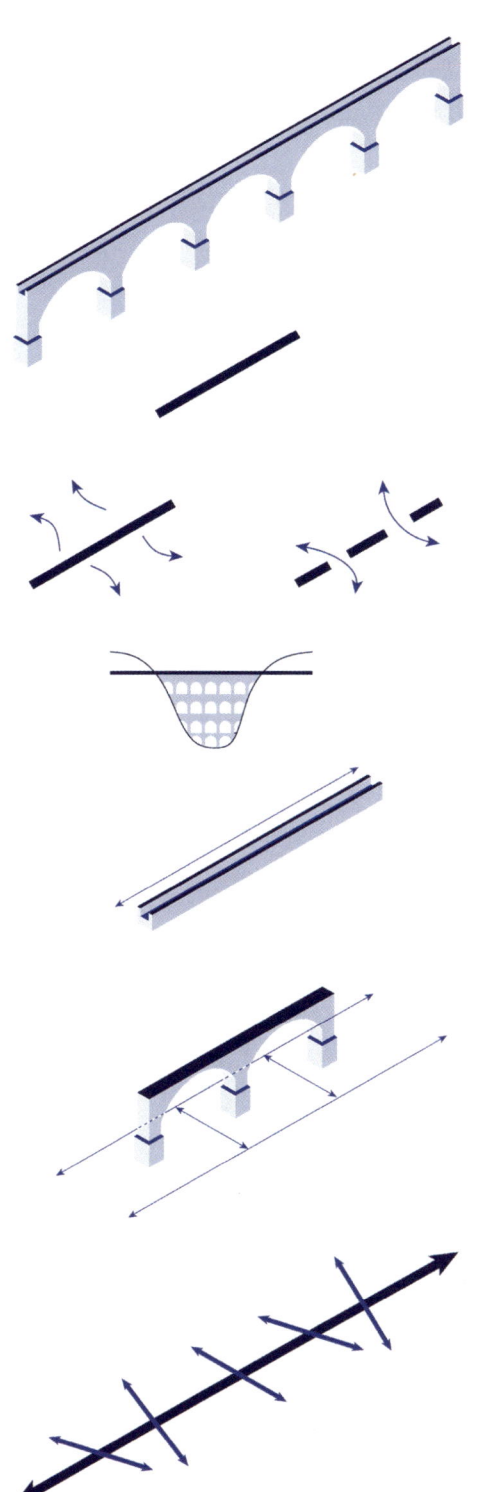

This network has been traced by strategic roads that accompany the identified avenues and cross transversely the urban edges in order to connect the fragmented areas. Finally, the strategy proposes to generate pedestrian bridges over the railway in order to connect the northern sector with the project area.

### New infrastructure

The project is a linear park with infrastructures for public use that gives it a historical and cultural character. This linear park expands towards the city through the road networks and the elevated paths over the railroad. The strategically generated paths are redesigned to favor pedestrian mobility with wide sidewalks, trees, seating areas and bicycle paths, creating a pleasant atmosphere on a human scale. The proposed volumes accompany the central path, presenting a break that directs and guides the path of people in their journey.

The infrastructure of the northern area dialogues directly with the sectors in front of the railway through the elevated walkways, while the lower volume dialogues with the existing infrastructure and generates public and semi-public spaces.

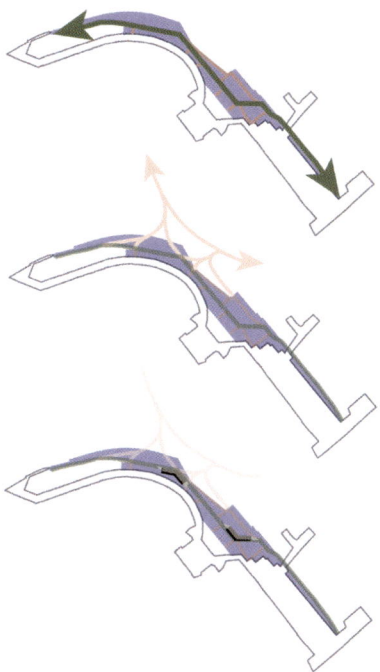

Concept schemes and project strategies.
Author: Angel Salinas.

Abstract view of the project buildings and functional scheme.
Author: Angel Salinas.

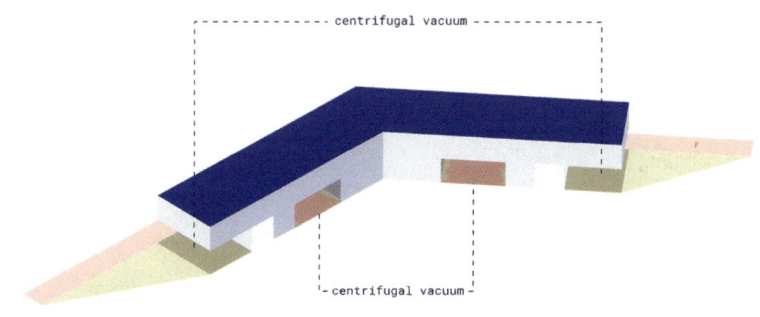

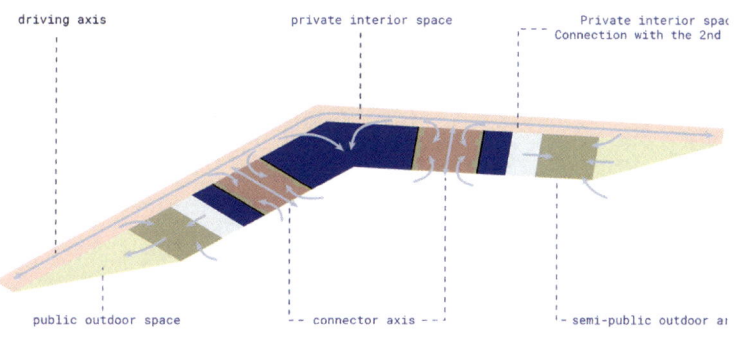

## Postface
# LANDSCAPE IS WHAT WE ARE IN
## *Carmen Andriani*

Landscape is an artifice, a cultural inheritance and an image of the mind that everyone elaborates on their own. It changes according to the field of perception and the orientation of the observer; it is irreducible to a univocal cartographic map, even if it remains a necessary step to the conventional representation. Franco Farinelli writes: "...unlike the place, the landscape is not made up of things but is only a way of seeing and representing the things of the world"[1] and, again, by André Corboz we read: "This landscape that I look at, disappears if I close my eyes and what you see from the same point is in any case different from what I see".[2]
There are three conditions for the epiphany of the *mental image* to occur: the observing subject, the observed object and the width of the horizon that is usually seen from above. The first is based on grandeur, the second one on scientific verifiability, a necessary condition for expressing a synthetic judgment and adjusting it according to the expression that we know best. In all cases, the observer is *no different* from the landscape in which he is immersed: what he sees is no longer the background of the representation but it is itself representation. Perceptions transcribed into painting, photography, notebook drawings or multimedia manipulations are the proof. The reversal of the meaning that Caspar David Friederich puts into effect in the *Wanderer above the Mist* in 1818 is powerful and revolutionary. The traveler, pictured from behind, contemplates, from the top of a peak, the sea of fog in front of him. The landscape, spreaded with the sentiment of the beholder, becomes the autonomous subject of the representation. It is not the restitution of a real landscape but the representation of the emotions that it creates inside the traveler and that involves the viewer, who, in turn, looks at the painting. In this case, the landscape becomes the *medium* that links what the artist sees *in front of* him to what he sees *inside himself*. The experience of the landscape deeply involves the person who looks at it, who lives in it and who crosses it. It could be said that the subject is an essential part of a spatiality that contributes to shaping it and by which it is molded, into a relationship of mutual necessity. The body is its reference and measure. It is the intermediary of the relationships between different focal points. This adds up another important element: the space of relationship, in other words, the perceptive, tactile and emotional relationship that is established between several parts held together through the void.
The dynamic tension that is generated between the objects produces a continuous variation of perception by shifting the perspective focus from the single individuality to the entire field of action and relationship. If on the one hand the void takes on an important aesthetic value, on the other the free links of the body into space generate forms: the changeability and unpredictability of these gestures produce continuous modifications to the point of metamorphosis.[3]
So what is the landscape if not "what is left inside the memory after you stop looking?" Gilles Clément writes in *Gardens, Landscape and Natural Genius* (2012). The concept, thus expressed, is inclusive, subjective, irreducible to a universal definition, to a univocal map or to a preordained dimensional scale. From a blade of grass to the vastness of the globe, *everything is landscape*. It is the condition of the contemporary city, it is the nature of the *hybrid landscapes* inside which we are immersed. Paraphrasing Joseph Rykwert therefore, «landscape is what we are in»[4].

1. Farinelli, F. (2003). Geografia, un'introduzione ai modelli del mondo. Torino: Einaudi. 2. Corboz, A. (1998). Ordine sparso. Saggi sull'arte, il metodo, la città, il territorio. (Viganò P., ed.). Milano: Franco Angeli. 3. Andriani, C. La Bellezza è un fraintendimento in Carpenzano, O., Nencini D., Raitano, M. (2018, eds.) Architettura in Italia. I valori e la bellezza, DiAP PRINT / Teorie. Macerata: Quodlibet. 4. The title is deliberately similar to the title of Joseph Rykwert's essay «Heritage is what we are in» (transl. «Il Patrimonio è ciò entro cui siamo»), preface to Andriani, C. (2010, ed.). Il Patrimonio e l'Abitare. Roma: Donzelli.

## SELECTED BIBLIOGRAPHY

Alehashemi, A., **Mansouri, S. A., Barati**, N. (2015), Urban Infrastructures and the Necessity of Changing Their Definition and Planning Landscape Infrastructure; a New Concept for Urban Infrastructures in 21st Century, *The Scientific Journal of NAZAR research center (Nrc) for Art, Architecture & Urbanism* [www.researchgate.net/publication/316493793_Landscape_infrastructure_A_new_concept_for_urban_infrastructures_in_21st_century].

**Allen** S. (1999). *Points + Lines: Diagrams and Projects for the City*, New York: Princeton Architectural Press.

**Andriani** C. (2016), "Reti minori ed Entroterra", in Bocchi R., Fabian L., Munarin S. (eds.), *Re-Cycle Italy. Atlante*, Siracusa: letteraventidue Edizioni, pp. 56–59.

**Andriani** C. (2014), "Ripensare l'infrastruttura. Note sul sistema ferroviario dismesso", in Cozza C., Valente I. (eds.), *La freccia del tempo*, Pearson Turin: Italia Edizioni, pp. 32–34.

**Bélanger** P. (2017). *Landscape as Infrastructure*. London: Routledge.

**Benedict**, M.A., **McMahon**, E.T. (2006), *Green Infrastructure: Linking Landscapes and Communities*, Arlington, Virginia: The Conservation Fund.

**Biehl**, D. (1991) *The Role of Infrastructure in Regional Development*, in Vickerman, R.W., Eds., Infrastructure and Regional Development, Pion, London, 3-9.

**Brenner** N., **Schmid** C. (2012). "Planetary urbanization", in Gandy M. *Urban Constellations*, Berlin: Jovis, pp. 10–14.

**Brunini**, C., **Paradisi**, F. (2006), *Le infrastrutture in Italia. Un'analisi provinciale della dotazione e della funzionalità*, vol.7 [www.istat.it/it/files/2011/03/UAN0146354InformazioniN7_Infrastrutture_in_Italia.pdf].

**Castells** M. (2008), La nascita della società in rete, UBE Paperback, Milan, pp. 435–491.

**Corner** J. (2006), "Terra fluxus", in Waldheim C. (ed.), *The Landscape Urbanism Reader*, New York: Princeton Architectural Press, pp. 21–33.

**Cozza** C., **Valente** I. (2014). *La Freccia del Tempo. Ricerche e progetti di architettura delle infrastrutture*, Milano: Pearson.

**Desimini** J., **Waldheim** C. (2016). *Cartographic Grounds. Projecting the Landscape Imaginary*, New York: Princeton Architectural Press.

**Doherty** G., **Waldheim** C. (eds.) (2015). *Is Landscape...? Essays on the Identity of Landscape*, Oxford: Routledge.

**Facchinelli** L. (2018). "Scali ferroviari, da infrastrutture di trasporto ad aree urbane", in *Trasporti&Cultura*, n. 52, pp. 5–7.

**Edwards**, P. N. (2003), *Infrastructure and modernity: force, time, and social organization in the history of sociotechnical systems*, in P. B. Thomas J. Misa, and Andrew Feenberg (Ed.), Modernity and Technology. Massachusetts: Massachusetts Institute of Technology.

European Commission, Bruxelles, 2013. *Infrastrutture verdi – Rafforzare il capitale naturale in Europa*. [www.eur-lex.europa.eu/resource.html?uri=cellar:d41348f2-01d5-4abe-b817-4c73e6f1b2df.0005.03/DOC_1&format=PDF]

European Commission, Science for Environment Policy, *In-depth Reports. The Multifunctionality of Green Infrastructure*, March 2012. [www.ec.europa.eu/environment/nature/ecosystems/docs/Green_Infrastructure.pdf].

European Environment Agency (EEA), *Green Infrastructure and territorial cohesion. The concept of green infrastructure and its integration into policies using monitoring systems*. EEA Technical report No 18/2011.

**Farinelli** F. (1991). "L'arguzia del paesaggio", in *Casabella*, n. 575-576, pp. 10–12.

**Farinelli** F. (2003). *Geografia. Un'introduzione ai modelli del mondo*. Bologna: Piccola Biblioteca Einaudi.

**Farinelli** F. (2009). *La crisi della ragione cartografica*, Bologna: Piccola Biblioteca Einaudi.

**Guallart** V., **Gausa** M., **Muller** W. (2000). *Diccionario Metapolis*

Arquitectura Avanzada, Barcelona: Actar.

**Hansen Niles** M. (1965), *The structure and determinants of local public investment expenditures*, Cambridge, Mass: MIT Press, ISSN 0034-6535, Vol. 47.1965, 2, p. 150-162.

**Insolera** I. (1962). *Roma moderna*, Torino: Einaudi.

**Kipar** A. (2010). "Infrastrutture e paesaggio", in *Ce.S.E.T.*, *Atti del XXXIX Incontro di Studio*, Firenze, pp. 47-53.

**Jakob** M. (2009). *Il paesaggio*. Bologna: Il Mulino, Universale Paperbacks.

**Jakob** M. (2005). *Paesaggio e letteratura*. Firenze: Olschki.

**LAND, Landscape Architecture Nature Development**, *Adaptive Design Research* [www.landsrl.com/adaptivedesign].

**Lynch** K. (1964). Ceccarelli P. (ed.). *L'immagine della città*. Milano: Marsilio Editore.

**Mattioli** V. (2019). *Remoria. La città invertita*. Roma: Minimum Fax.

**McMahon**, M. B. E. (2002). "Green Infrastructure: Smart Conservation for the 21st Century", in *Renewable Resources Journal*, (20), pp. 2-17.

**Montedoro** L. (2018). "Milano, scali ferroviari e trasformazione della città", in *Trasporti&Cultura*, n. 52, pp. 35-45.

**Montuori** L. (2019). "Roma, verso il progetto urbano delle stazioni", in *EcoWeb Town*, n. 20, pp. 140-153.

**Pagnotta** G. (2012). *Dentro Roma. Storia del trasporto pubblico nella capitale (1900-1945)*. Roma: Donzelli. Quilici V. (2007). *Roma, Capitale senza centro*, Roma: Officina.

**Pietrolucci** M. (2012). *La città del Grande Raccordo Anulare*, Roma: Gangemi.

**Recalcati** S., **Fraticelli** C. (2018). "La rigenerazione degli scali ferroviari in Europa. Tre esperienze a confronto", in *Trasporti&Cultura*, n. 52, pp. 89-99.

**Science for Environment Policy**, *In-depth Reports | The Multifunctionality of Green Infrastructure* | March 2012. [www.ec.europa.eu/environment/nature/ecosystems/docs/Green_Infrastructure.pdf].

**Snyder** S. N., **Wall** A. (1998). "Emerging landscapes of movement and logistics", in *Architectural Design Profile*, n. 134, pp. 16-21.

**Toni**, F. (2014), *Le infrastrutture verdi i servizi ecosistemici e la green economy*, Fondazione per lo sviluppo sostenibile. [www.comitatoscientifico.org/temi%20CG/documents/MATTM%20IV%20310314.pdf].

**Waldheim** C. (2016). *Landscape as Urbanism. A General Theory*, Princeton and Oxford: Princeton University Press.

**Waldheim** C. (ed.)(2006). *The Landscape Urbanism Reader*, New York: Princeton Architectural Press.

**Williams**, R. (2012), "Systems & Strategies for Contemporary Urbanization", Landscape Infrastructure, Piper Auditorium, Cambridge, MA.

## WEB REFERENCES

C40 - Reinventig Cities
[www.c40reinventingcities.org]

FS Sistemi Urbani, Progetto Roma - Roma Tuscolana
[www.fssistemiurbani.it/content/fssistemiurbani/it/opportunita-di-investimento/roma/tuscolana.html]

FS Sistemi Urbani, Progetto Milano - Scali Ferroviari Milano
[www.fssistemiurbani.it/content/fssistemiurbani/it/scali-milano.html]

SCALI FERROVIARI - BENCHMARKING DI RIGENERAZIONI URBANE DI SUCCESSO SU AREE FERROVIARIE DISMESSE Arup Italia srl per FS Sistemi Urbani Milano, 2016
[www.fssistemiurbani.it/content/dam/fs-sistemi-urbani/scali-milano/Arup_Report-Scali-Ferroviari_150916_RED.pdf]

Deltawerken online
[www.deltawerken.com]

## CONTRIBUTORS

**Carmen Andriani**
Architect, Full Professor of Architecture and Urban Design at dAD, UniGe. Her scientific research includes the relationship between infrastructure, landscape and heritage. Since 2014, she directs the Integrated Studio Coastal Design Lab at dAD. She has been a Visiting Professor in architectural schools such as Tongji University and Florida International University. She publishes and lectures worldwide (Canada, Spain, Uruguay).

**Manuel Gausa**
Architect, Full Professor of Urban Design and Planning since 2017 at dAD, UniGe. Coordinator of the ADD, PhD in Architecture and Design at the Department of Architecture and Design (DAD-UNIGE - University of Genoa) directs the GIC Lab, Research laboratory in architecture, urbanism and landscape. Co-founder of IAAC, Institut d'Arquitectura Avancada de Catalunya (Spain).

**Nicola V. Canessa**
Architect, PhD and Assistant Professor of Urban Design and Planning at dAD, UniGe. His research interests include "The Mediterranean City" as analysis of the changes taking place on the Mediterranean coasts and the factors that affect the intercontinental inhabit the coastal territory. Currently Canessa is part of the GicLab research group in Genoa.

**Francesco Garofalo**
Born in France in 1983, he moved to the Netherlands in 2008 where he founded Openfabric in 2011, an international urban/rural design practice based in Rotterdam and Milan, with projects ranging from small installations up to the large geographical scale of regional strategies. He is Visiting Professor of Landscape Architecture Master at Politecnico di Milano and Visiting Lecturer at London Metropolitan University.

**Luigi Mandraccio**
Architect and PhD at dAD, UniGe, 2021. His main research topic is the Big Science structures. This topic is developed within the general theme of industrial architecture, considering wider interactions with contexts and people. Author of many essays and articles, Mandraccio is involved in several editorial projects.

**Matilde Pitanti**
Architect and PhD at dAD, UniGe. Her research investigates the potential of resilient territorial strategies, in response to hydrogeological and flooding problems in urban riverfront, as urban regeneration. She studied at the Technische Universität in Wien, through the Erasmus Plus program. Currently Pitanti is part of the GicLab research group in Genoa.

**Flavia Rossi**
Photographer based in Rome and Milan. She graduated at the Architecture Faculty of the University of Rome La Sapienza in 2016, with a thesis on the Aesthetics of the Landscape. In 2017 she graduated at the Master Iuav in Photography. In 2019 she won the Atlante dell'architettura contemporanea Competition, supported by MUFOCO in collaboration with Triennale and Direzione Generale Creatività Contemporanea.

**Davide Servente**
Architect, PhD, Assistant Professor in Architectural and Urban Composition at at dAD, UniGe. He founded the multidisciplinary research collective ICAR65. His research focuses on the relationship between public art, urban spaces, ordinariness in architecture and strategies for coastal heritage reuse. In 2020, he published for Sagep the books "Il Museo PopUp. Arte pubblica e spazio urbano" and "Abitare nel Tempo. Venti ville del Novecento" (with A. Canevari).

**Emanuele Sommariva**
Architect, PhD, Assistant Professor of Urban Design at dAD, UniGe. He has been Researcher at LUH Hannover (2012-2020). Visiting scholar at TUMünchen and Universiteit Antwerpen. His research interests include Urban Recycle and Resilience, Landscape Urbanism, Food-Design-Territory. Scientific Manager of EU project: Creative Food Cycles (2018-20) Regio Branding (2017-19) COST Urban Allotment EU (2012-16).

## AUTHORS

### Beatrice Moretti

PhD Architect, Lecturer and Research Fellow from dAD - Department Architecture and Design of Genoa (UniGe). Her research research has focused on contemporary port cities, port clusters and the architecture of port cities. Currently, Moretti is developing new lines of research on cross-cutting themes between architecture and urbanism, such as university systems conceived as multi-scalar landscape infrastructures and high-density residential complexes. From 2010 to 2015, she was Research Fellow at Genoa Urban Lab and at Genoa Port Authority. Since 2015, she is Teaching Assistant within the Integrated Studio *Coastal Design Lab*, led by Professor C. Andriani. In 2018, she co-founds *caarpa*, a collective project that combines architecture and landscape; in the same year, she has been Guest Researcher at Delft University of Technology (NED). In 2019, she was Lecturer in Landscape Urbanism at dAD (UniGe) and, from 2021, at DICCA (UniGe) in Architectural Composition. Since 2020, she collaborates in teaching activities at DAStU (POLIMI) on Landscape and Garden Design topics and, on behalf of dAD, in the International Architectural Seminar "Villard". Since 2021, she is co-curator of the events *BOOK TALKS* in collaboration with dAD and the Politecnico di Torino and she is part of the interdisciplinary research group *PortCityFutures* (Leiden-Delft-Rotterdam Universities, NED). Moretti lectures regularly in Europe and beyond. She is author of the books "Un colle, un transatlantico, un nome. Tre storie sul porto di Genova" (Sagep, 2018) and "Beyond the Port City. The Condition of Portuality and the Threshold Concept" (JOVIS, 2020).

### Giorgia Tucci

PhD Architect, Lecturer and Research Fellow from dAD - Department Architecture and Design of Genoa (UniGe). Her PhD research rethinks the identity of rural coastal cities in the Mediterranean area - MedCoast AgroCities - through territorial strategies (economic, energy, environmental, cultural and social), integrated with the application of new technological systems and innovative planning approaches. Founder of the website platform *agrocities.com*. In 2017-2018, she carried out as Guest Researcher a doctoral research period at the ETS, Escuela de Arquitectura in Málaga. In 2017 she was Visiting Lecturer at Universidad de Málaga, UMA-eAM' (Malaga, Spain), in 2019 at Leibniz Universität Hannover LUH (Hannover, Germany) and in 2020 at Institute of Advanced Architecture of Catalonia IAAC (barcelona, Spain). Since 2015 she is coordinator of the GicLab research group at dAD-UniGe, founded by Prof. Manuel Gausa. Since 2019 she is Research Fellow in EU Project Creative Europe Programme – Creative Food Cycles Project - and works actively in the management and writing of European projects. She was lecturer in Urban planning for the landscape (2019-2020) and Urban systems and new technologies (2021-2022) at dAD. She is author of the books "Albenga GlassCity. From the GlassCity to the GreenCity" (ListLab, 2018) and "MedCoast AgroCities. New operational strategies for the development of the Mediterranean agro-urban areas" (ListLab, 2019) and continues her scientific production within the GicLab research group in Genoa.

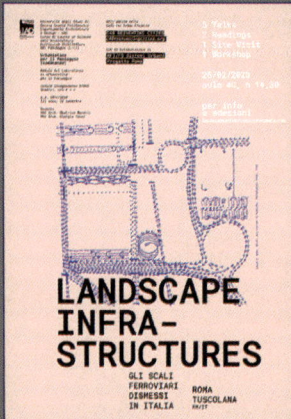
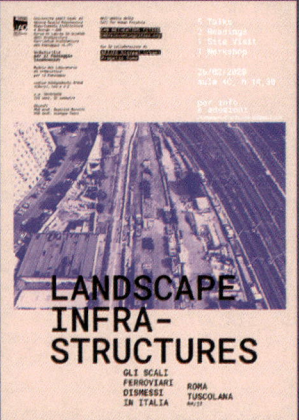
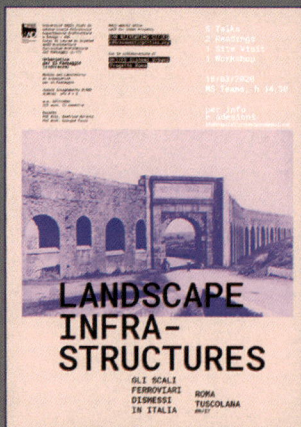

**A Landscape Infra-structures Research**
ROMA TUSCOLANA Pilot Project

**Authors**
Beatrice Moretti
Giorgia Tucci

**Editorial Director of LIStLab**
Alessandro Martinelli

**Published by**
ListLab
*info@listlab.eu*
*listlab.eu*

**Art Director & Production**
Blacklist Creative, BCN
*blacklist-creative.com*

**ISBN** 9788832080643

series

**Printed and bound**
in the European Union, 2022

**All rights reserved**
© of ListLab edition;
© of the author's texts;
© of the author's images.

No part of this book may be reproduced, stored in a retrieval system, or transmitted in any form or by any means, including electronic, mechanical, photocopying, microfilming, recording or otherwise without written permission, except in the case of brief quotations embodied in critical articles and reviews. Every reasonable effort has been made to contact the rightful copyright owners after their academic course involvement. We apologise for any inadvertent errors or omissions.

**Sales, Marketing & Distribution**
*distribution@listlab.eu*
*listlab.eu/en/distributori-promotori-ditributors-promotors/*

For more information concerning Listlab's Scientific Boards please visit the webpage:
*https://www.listlab.eu/en/board-comitati-listlab/*

**ListLab** was established in 2007 and has elaborated on the idea of an international editorial laboratory with a multidisciplinary approach to architecture, planning, arts, photography, and design. List Group, found in 2021, aims at creating networks and promoting debates and cultural exchange, but also organize events from which new knowledge about architecture, cities, and landscape can develop. Today, **List Group** is composed of **ListLab**, the publishing house, **Blacklist**, the graphic design studio, **Instaura**, the informational weblog, and **Us/Them/Yours**, a creative agency that aims at a multimedia approach to information.

This publication has been possible thanks to a grant from the dAD, Department Architecture and Design, Polytechnic School, UniGe.

**UniGe | DAD**

## ACKNOWLEDGMENTS

### LANDSCAPE INFRASTRUCTURES STUDIO
Course "Urbanistica per il paesaggio" (Module of "Laboratorio di urbanistica per il paesaggio"), University of Genoa (ITA), Department Architecture and Design – dAD, Course in Architectural Sciences – Landscape Architecture curriculum (year 2019/2020)

### LECTURERS
PhD. Arch. Beatrice Moretti
PhD. Arch. Giorgia Tucci

### STUDENTS
Roberta Cereghino, Andrea Derni, Cecilia De Mattia, Andrea Oriati, Elisa Pizzorno, Cassi Ranaivoson, Angel Salinas Escandon, Arianna Viola.

### THANKS TO
FS Sistemi Urbani S.r.l.
Area Centro, Progetto Roma
*www.fssistemiurbani.it*

### WITHIN THE FRAMEWORK OF
"C40 Reinventing Cities"
Call for Urban Projects
*www.c40reinventingcities.org*

## CREDITS

### PROJECTS
*Texts and graphic elaborations*
Andrea Derni [pp. 180-185]
Roberta Cereghino [pp. 186-191]
Andrea Oriati, Elisa Pizzorno, Arianna Viola [pp. 196-203]
Cecilia De Mattia, Cassi Ranaivoson [pp. 204-207]
Angel Salinas Escandon [pp. 208-213]

### PHOTOS
© Flavia Rossi, *Osservatorio sul Paesaggio*, 2021.
[cover, pp. 6-7; 14-15;46-47; 80-81; 144-145; 152-153; 156-157; 176-177; 192-193; 214-215]

**Where not explicitly stated, the texts are equally attributable to Beatrice Moretti and Giorgia Tucci.**